Data Analytics, Computational Statistics, and Operations Research for Engineers

Data Analytics, Computational Statistics, and Operations Research for Engineers

Methodologies and Applications

Edited by

Debabrata Samanta, SK Hafizul Islam,
Naveen Chilamkurti, and Mohammad Hammoudeh

CRC Press
Taylor & Francis Group
Boca Raton London

CRC Press is an imprint of the
Taylor & Francis Group, an **informa** business

First edition published 2022
by CRC Press
6000 Broken Sound Parkway NW, Suite 300, Boca Raton, FL 33487–2742

and by CRC Press
2 Park Square, Milton Park, Abingdon, Oxon, OX14 4RN

CRC Press is an imprint of Taylor & Francis Group, LLC

Library of Congress Cataloging-in-Publication Data
Names: Samanta, Debabrata, 1987– editor.
Title: Data analytics, computational statistics, and operations research for engineers : methodologies and applications / edited by Debabrata Samanta, SK Hafizul Islam, Naveen Chilamkurti, and Mohammad Hammoudeh.
Description: First edition. | Boca Raton, FL : CRC Press, [2022] | Includes bibliographical references and index.
Identifiers: LCCN 2021046713 (print) | LCCN 2021046714 (ebook) | ISBN 9780367715113 (hbk) | ISBN 9780367715120 (pbk) | ISBN 9781003152392 (ebk)
Subjects: LCSH: Engineering—Data processing.
Classification: LCC TA345 .D28 2022 (print) | LCC TA345 (ebook) | DDC 620.001/51—dc23/eng/20211118
LC record available at https://lccn.loc.gov/2021046713
LC ebook record available at https://lccn.loc.gov/2021046714

ISBN: 978-0-367-71511-3 (hbk)
ISBN: 978-0-367-71512-0 (pbk)
ISBN: 978-1-003-15239-2 (ebk)

DOI: 10.1201/9781003152392

Typeset in Times
by Apex CoVantage, LLC

To my parents Mr. Dulal Chandra Samanta, Mrs. Ambujini Samanta, my elder sister Mrs. Tanusree Samanta, and daughter Ms. Aditri Samanta

Dr. Debabrata Samanta

To my son Mr. Enayat Rabbi

Dr. SK Hafizul Islam

Contents

Preface

Data Analytics, Computational Statistics, and Operations Research for Engineers: Methodologies and Applications focuses on devising efficient methodologies to obtain quantitative solutions for problems that are devised quantitatively. This subject brings together computational capability and statistical advanced thought processes together to solve some of the issues efficiently. Some of the interesting areas of the subject are optimization techniques in the process of statistical inferencing, algorithms of expectation maximization, and Monte Carlo simulation to name a few. Advances in simulation and graphical analysis also add to the pace of the mathematical analytics field. Computational statistics play a crucial role in financial applications, particularly risk management and derivative pricing, and biological applications include bioinformatics and computational biology and computer network security applications that touch the lives of people. With high-impacting areas such as these, it becomes essential to dig deeper into the subject and explore the key areas and their progress in the recent past.

The book includes the present trends and future directions of various computational statistical methods and applications in various domains. This book is divided into three sections. In the first part, the book discusses various statistical methods for computing, including arithmetic, numerical methods, software, simulation techniques, optimization algorithms, statistical graphics, and so on. Besides, this part also includes the role of modern computer and linear algebra in computing and various transformation functions in restructuring the problem statements. In the second part, this book discusses various computation-intensive tactical, knowledge inference techniques, Bayesian analysis tools, survival analysis models, and impact of data mining on the computing. The last part of the book focuses on the applications of computational statics in various fields, such as finance and economics, human biology, clinical research, and network security.

Chapter 1 explores that over the past 20 years, how hyperspectral imaging has been actively researched as an emerging, promising approach for assessing the quality and preservation of horticultural and agricultural products. This technology grew out of remote sensing and combines machine vision and point spectroscopy to deliver superior image segmentation for defect and contamination identification. We have integrated spatial and spectral information into hyperspectral imaging algorithms in this chapter. It can provide useful information on both outward physical and internal chemical features of agricultural and food goods in a quick and nondestructive manner. This chapter summarized the most important aspects of the applicability of hyperspectral imaging in precision agriculture. This research examines hyperspectral imaging's present and historical applications in agriculture, including the following categorization: chromatic, climatic, and convergence. The goal of this analysis is to identify prospective work on significant issues and promote implemented end-user solutions in order to meet the existing global sustainability goals.

Chapter 2 discusses about the emergence of the COVID-19 virus as one of the most widely perceived diseases. Based on three accessible datasets, the research achieved several deep learning convolutional networks with prior training strategies for categorizing COVID-19 chest X-ray pictures into three classes: COVID-19 positive and negative. Our dataset contained 980 X-ray pictures from COVID-19 patients, and we tried and tested several approaches to get the best results possible. When we had an unbalanced dataset, we presented certain pretraining strategies in our research that could assist the network to learn better (fewer cases of COVID-19 along with more cases of other classes). We also suggested a modified Convolutional Neural Network in which activation is achieved by applying a filter to an input. The location and strength of a recognized feature in an input, such as an image concatenation of neural networks and VGGNet networks, are shown by a map of activation termed as a feature map, which is created by repeatedly applying the same filter to input. By combining several features collected by two robust networks, this network was able to attain the highest accuracy. We tested our network on 1,960 photos to see what level of accuracy it could achieve in real-world situations. The suggested network's average detection accuracy for COVID-19 cases was 96.60%, and its average performance and accuracy for provided classes was 61.6%. The suggested work completed a comparative research with factors such as accuracy, time complexity, and high performance, resulting in significant cost savings. The existing solution is better in terms of computational cost and functioning with a little amount of training data.

Chapter 3 emphasizes that with the spread of democracy around the world, election frauds such as vote buying, booth rigging, and booth polling have become increasingly common. Three decades ago, the neo-concept of an electronic voting system was born. In contrast to the manual voting system with paper ballots, it uses voting computers connected to a public network to conduct the election process. Regardless, it has been largely ineffective in preventing an electoral fraud. As a result, with the constantly increasing need for a secure and transparent E-voting system, it has become imperative to make some radical improvements to the current E-voting system, making it more accessible to the general public while also offering consistent voting results. This chapter discusses blockchain technology as a critical service provider in modern society as well as the development and deployment of a distributed ledger technology-based application to improve the current electronic voting method. The chapter will focus on blockchain technology's decentralized design and its function in spreading digital information synchronously throughout a network to ensure the integrity of polled and un-polled votes across all EVMs in the network. At the end, the importance of a blockchain-based E-voting system in overcoming the limits of the current E-voting system in order to conduct a cost-effective, transparent, and secure nationwide election will be examined.

In Chapter 4, it is discussed that in 2020 the global health industry faced the most serious public health danger in a decade. While the COVID-19 epidemic reminds us of situations like SARS and Ebola, the breadth and scope of this pandemic have taken us into a new territory, posing critical issues about how we should respond to similar disasters. The rapid spread of the virus has left users, businesses, and governments reeling. With their complete attention and resources, researchers focused on modeling the likely propagation of the virus and numerous processes that may impact the suppression of the spread of the virus in their hunt for remedies. So, how have we attempted to control COVID-19? In the health sector, there was a rising reliance on technology to give prompt solutions. Artificial Intelligence, or AI, has already provided solutions to a number of everyday difficulties we have encountered. Intelligent solutions have also come to our rescue to make our lives much easier by assisting in learning more effectively by enrolling in personalized AI-enabled teaching assistants to assist us in being intelligent in our homes or at work. AI has now provided solutions for the coronavirus epidemic as well. Smart solutions, in conjunction with continuous research, help explain how the virus evolves and how it can be dealt with more swiftly and cheaply. AI companies all over the world have been battling to develop new solutions to combat the new coronavirus and limit the damage it causes. In other words, the direct expansion of cases and their health records has been a key source of information and awareness. Various data storage technologies are used to store a large number of data from these scenarios right away. These data are utilized to undertake virus research and development as well as pandemic preparedness and response, as well as action against the virus and its consequences. Big data is a cutting-edge technology that can digitally store a large number of these patients' records. It aids in the computational examination of patterns, trends, relationships, and differences. It may also aid in the discovery of new information on the disease's spread and control. By providing extensive data-capturing capacity, big data may be effectively leveraged to limit the risk of this virus spreading. This chapter highlights a few key techniques for AI to be effective as well as a few COVID-19 big data applications.

Chapter 5 discusses that blockchain and AI have matured to the point where they are now driving advancements in almost every industry. AI refers to computers that are programmed to carry out clever tasks that have been developed by humans. Blockchain is a decentralized network of computers, which collects and saves data to demonstrate an orderly sequence of events on a simple and permanent record system. AI and blockchain are becoming an absolutely wonderful combo that is rapidly improving. It is undeniable that AI and blockchain concepts are gaining traction at a rapid pace. Both ideas have a high level of mechanical complexity as well as multidimensional commercial suggestions. In any event, a common misconception about the blockchain concept is that "a blockchain is decentralised, and so nobody controls it." In any case, a group of center designers is still recognized with the essential development of a blockchain system. Take, for example, smart agreement, which is nothing more than a collection of codes (or capabilities) and information (or states) changed and sent on a blockchain (say, Ethereum) by a group of human software developers. As a result, it is less likely to be free of escape clauses and flaws, which is surprising. They are realized regardless of the industry in which they are realized. With blockchain and AI, everything from grocery store network coordination and medical data interchange to media sovereignty and monetary security is being reimagined. Because both deal with information and value, a blockchain–AI collaboration is unavoidable. Blockchain allows for the secure storage and sharing of information and other valuable assets. AI can investigate and create knowledge from data in order to generate esteem.

In Chapter 6, it is discussed that in commerce, research, business, and society, big data analytics has made revolutionary discoveries. Companies in the social networking domain include Facebook, Twitter, and LinkedIn. Scientific and technological advancements in the field of data analytics have aided in the economically extraction of value from enormous volumes of a variety of data to empower people in their professional and personal life. The rise of big data has had a significant impact on man's work, thought, and decision-making over the years. The ability to extract data through data analytics is the X-factor for getting the most out of this. Machine learning (ML) is at its pinnacle because of its ability to provide meaningful data insights, predictions, and decisions, with recent studies highlighting some promising learning methods such as deep learning, representation learning, distributed and transfer learning, parallel learning, kernel-based learning, and active learning as promising learning methods. Big data analytics aids in the discovery of diverse market trends, correlations, hidden patterns, and client preferences, all of which aid firms in making more informed decisions. With the help of a variety of decision-making algorithms, ML aids in the acceleration of this process. Big data and ML together can identify incoming data and translate it into insights for a variety of business needs. This book focuses on how the interconnectedness of big data and ML might help solve a variety of problems that are now being faced. Furthermore, a research foundation that encourages the development of new innovative techniques in the field of ML with big data.

In Chapter 7, Internet of Things (IoT) has been praised by both large organizations and private clients throughout the years. However, a widespread adoption of this new technology raises a slew of security concerns. Simply put, anything connected to the Internet poses a potential security risk to the network as a whole. According to a poll of more than 100 IoT industry professionals conducted at the Internet of Things World conference in San Francisco, the top two areas of technology adoption concerns are security and implementation. According to the 2020 Unit 42 IoT Threat Survey, 98% of IoT computer traffic is unencrypted, leaving the vast majority of personal network data and confidentiality vulnerable to cyber attacks. In addition, programmers revere 57% of IoT devices since they admire medium severity attacks. Given the preferences and exponentially growing popularity of IoT, it is worth rethinking how we handle the security of IoT devices and networks. One option to improve security and dependability in an IoT environment is to employ blockchain technology. Associated devices receive greater security and are less susceptible to malware and other attacks by decentralizing blockchain IoT networks and removing single failure points. Other advantages of a decentralized IoT architecture include more independent operations and lower network and foundation maintenance expenses. This chapter provides an overview of blockchain principles, with a focus on how blockchain technology can be used to protect IoT.

In Chapter 8, it is discussed that the importance of digital security in today's society cannot be overstated. Detecting the type of attack and recognizing trends in staged methods are the first steps toward establishing security. Phishing is a word used to describe these types of attacks. To be more accurate with the definition of phishing, it may be described as a social engineering attack involving complex attack vectors that can be utilized to obtain sensitive information from a victim. In phishing, the attacker poses as a trustworthy entity and employs a variety of strategies to gain the victim's trust before persuading the victim to provide the information needed to commit the fraud. This type of assault can be carried out by sending malicious emails carrying malware that, once installed on the victim's system, can leak information. Such assaults can then propagate to other victims via the hijacked systems. Email phishing, spear phishing, whaling, smishing and vishing, and angler phishing are the several types of phishing. The chapter focuses on spear phishing as being the most dangerous of them. Spear phishing is a type of email or social media scam that targets a single person or organization. Spear phishing is an email or social media scam that targets a specific person or organization, usually one with a lot of power. In spear phishing, the attacker acquires as much information about the target as possible in order to personalize the assault for that specific victim. These attacks, on the other hand, frequently follow a similar pattern. As a result, spotting that pattern in every email or social media post might alert you to any potential threats. Using ML algorithms, this chapter focuses on discovering these trends and categorizing posts as prospective spear-phishing attempts. The categorization algorithms divide data into a set of decision classes based on its structure or unstructured nature. Support vector machine (SVM), decision tree, logistic regression, multinomial Naive Bayes, and K-nearest neighbors (KNNs) are examples of classification methods. Ensemble approaches are also utilized to improve the accuracy of the data classification. Ensemble approaches integrate many ML techniques to balance variance and bias or improve predictions in ML. These ensemble methods are further separated into sequential and parallel ensemble methods, such as AdaBoost, Stacking, and Random Forest. Both approaches were employed in this work to compare the accuracies of traditional ML algorithms and ensemble methodologies. Furthermore, the ensemble approaches outperform classic ML techniques in predicting Spear phishing in an email, according to our research.

Chapter 9 explores RFID as a new technology that can be used to track, identify, and track physical objects. In today's competitive business world, technology plays a critical role in improving business unit operational efficiency. The purpose of this study is to learn about effective RFID methods for increasing operational performance of selected medical stores in Odisha, India. Data was obtained from key officials of retail units, i.e., store manager, operations manager, procurement manager, etc. from medical stores affiliated with 15 private medical registered under the Government of Odisha using various statistical tools such as correlation and regression. The study finds that RFID is the most efficient technique and has a substantial impact on the operational performance of the Medical Store within the hospital as well as contributing to the body of knowledge and assisting hospital management practitioners in the healthcare industry.

Acknowledgments

We express our great pleasure, sincere thanks, and gratitude to the people who significantly helped, contributed, and supported to completion of this book. We are sincerely thankful to Dr. G.P. Biswas, Emeritus Fellow, Department of Computer Science and Engineering, Indian Institute of Technology (Indian School of Mines) Dhanbad, Jharkhand, India, for his encouragement, support, and guidance, advice, and suggestions to complete this book. Our sincere thanks are due to Dr. Siddhartha Bhattacharyya, Professor, Department of Computer Science and Engineering, CHRIST (Deemed to be University), Bengaluru, Karnataka, India, and Dr. Arup Kumar Pal, Associate Professor, Department of Computer Science and Engineering, Indian Institute of Technology (Indian School of Mines) Dhanbad, Jharkhand, India, for their continuous support, advice, and cordial guidance from the beginning to the completion of this book.

We would also like to express our honest appreciation to our colleagues at the Indian Institute of Information Technology Kalyani, and CHRIST (Deemed to be University), Bengaluru, Karnataka, India, for their guidance and support.

We also thank all the authors who have contributed some chapters to this book. This book would not have been possible without their contribution.

We are also very thankful to the reviewers for reviewing the book chapters. This book would not have been possible without their continuous support and commitment toward completing the chapters' review on time.

To complete this book, all the publishing team members at Taylor & Francis/CRC Press extended their kind cooperation, timely response, expert comments, guidance, and we are very thankful to them.

At the end, we sincerely express our special and heartfelt respect, gratitude, and gratefulness to our family members and parents for their endless supports and blessings.

Debabrata Samanta, PhD, MIEEE
Department of Computer Science
CHRIST (Deemed to be University)
Bengaluru, Karnataka 560029
Email: debabrata.samanta369@gmail.com

SK Hafizul Islam, PhD, SMIEEE
Department of Computer Science and Engineering
Indian Institute of Information Technology Kalyani, West Bengal 741235, India
Email: hafi786@gmail.com

Naveen Chilamkurti, PhD, SMIEEE
Professor and Head of Cybersecurity Discipline
Computer Science and IT
La Trobe University, Melbourne, Australia

Mohammad Hammoudeh, PhD, SMIEEE
Reader in Future Networks and Security
Leader of the CfACS IoT Lab
Department of Computing and Mathematics
Manchester Metropolitan University, Manchester, UK

Acknowledgments

Editor Biographies

Debabrata Samanta is presently working as an assistant professor, Department of Computer Science, CHRIST (Deemed to be University), Bangalore, India. He obtained his Bachelors in Physics (Honors), from Calcutta University, Kolkata, India. He obtained his MCA from the Academy of Technology, under WBUT, West Bengal. He obtained his PhD in computer science and engineering from National Institute of Technology, Durgapur, India, in the area of SAR Image Processing. He is keenly interested in Interdisciplinary Research & development and has experience spanning fields of SAR Image Analysis, Video surveillance, Heuristic algorithm for Image Classification, Deep Learning Framework for Detection and Classification, Blockchain, Statistical Modelling, Wireless Adhoc Network, Natural Language Processing (NLP), and V2I Communication. He has successfully completed six consultancy projects. He has received funding under International Travel Support Scheme in 2019 for attending conference in Thailand. He has received Travel Grant for speaker in Conference, Seminar, etc., for two years from July 2019. He is the owner of 20 patents (three designed Indian patents and two Australian patents granted, 15 Indian patents published) and two copyrights. He has authored and coauthored over 175 research papers in international journals (SCI/SCIE/ESCI/Scopus) and attended conferences including IEEE, Springer, and Elsevier Conference proceeding. He has received "Scholastic Award" at 2nd International conference on Computer Science and IT application, CSIT-2011, Delhi, India. He is a coauthor of 11 books and the coeditor of seven books, available for sale on Amazon and Flipkart. He has presented various papers at international conferences and received Best Paper awards. He is the author and coauthor of 20 book chapters. He also serves as acquisition editor for Springer, Wiley, CRC, Scrivener Publishing LLC, Beverly, USA, and Elsevier. He is a Professional IEEE Member, an Associate Life Member of Computer Society of India (CSI), and a Life Member of Indian Society for Technical Education (ISTE). He is a convener, keynote speaker, session chair, cochair, publicity chair, publication chair, advisory board, technical program committee member in many prestigious international and national conferences. He was invited as a speaker at several institutions.

SK Hafizul Islam received M.Sc. degree in applied mathematics from Vidyasagar University, Midnapore, India, in 2006, and M.Tech. degree in Computer Application and Ph.D. degree in Computer Science and Engineering in 2009 and 2013, respectively, from Indian Institute of Technology [IIT (ISM)] Dhanbad, Jharkhand, India, under the **INSPIRE Fellowship Ph.D. Program (funded by the Department of Science and Technology, Government of India)**. He is currently an Assistant Professor in the Department of Computer Science and Engineering, Indian Institute of Information Technology Kalyani (IIIT Kalyani), West Bengal, India. Before joining the IIIT Kalyani, he was an Assistant Professor with the Department of

Computer Science and Information Systems, Birla Institute of Technology and Science, Pilani (BITS Pilani), Rajasthan, India. He has more than eight years of teaching and 11 years of research experience. He has authored or coauthored 125 research papers in journals and conference proceedings of international reputes. His research interests include Cryptography, Information Security, Neural Cryptography, Lattice-based Cryptography, IoT & Blockchain Security, and Deep Learning.

Presently, he is editing four books for the publishers **Scrivener-Wiley, Elsevier**, and **CRC Press**. He is an **Associate Editor** for **"IEEE Journal of Biomedical and Health Informatics", "IEEE Transactions on Intelligent Transportation Systems", "IEEE Access", "International Journal of Communication Systems (Wiley)", "Telecommunication Systems (Springer)", "IET Wireless Sensor Systems", "Security and Privacy (Wiley)", "Array-Journal (Elsevier)", "Wireless Communications and Mobile Computing (Hindwai)", "Journal of Cloud Computing (Springer)", "Computer Communications (Elsevier)"**, and **Cyber Security and Applications (KeAi)**. He was the recipient of the **University Gold Medal**, the **S. D. Singha Memorial Endowment Gold Medal**, and the **Sabitri Parya Memorial Endowment Gold Medal** from Vidyasagar University, in 2006. He was also the recipient of the **University Gold Medal** from IIT(ISM) Dhanbad in 2009 and the **OPERA award** from BITS Pilani in 2015. He is a senior member of IEEE and a member of ACM.

Naveen Chilamkurti is currently a reader/associate professor and cybersecurity discipline head, Department of Computer Science and Computer Engineering, La Trobe University, Melbourne, VIC, Australia. He obtained his PhD degree from La Trobe University. He is also the Inaugural Editor-in-Chief for *International Journal of Wireless Networks and Broadband Technologies* launched in July 2011. He has published about 300 journal and conference papers. He has edited and authored six books with various publishers. His current research areas include Cybersecurity, IoT, anomaly detection in IoT, Internet of Medical Things, Wireless security, Federated Learning in IoT, wireless multimedia, wireless sensor networks, and so on. He currently serves on the editorial boards of several international journals. He is a senior member of IEEE. He is also an associate editor for IEEE, Wiley IJCS, SCN, Inderscience JETWI, and IJIPT.

Mohammad Hammoudeh received his MSc in advanced distributed systems in 2006 (University of Leicester, UK) and his PhD in wireless sensor networks in 2008 (University of Wolverhampton, UK). He was appointed as the inaugural head of the CfACS Internet of Things Lab in 2016. He was awarded a Readership in Future Networks and Security in 2019. He is an active disseminator of research to the academic community, industry, policy makers, and wider technology beneficiaries; he has over 75 refereed conference publications, 50 peer-reviewed journal publications, and is a successful editor of three books and many journal special issues. He continues to be an active researcher in the areas of cybersecurity, the Internet of things (IoT), and simulation and modeling of complex highly decentralized systems. Mohammad is engaged with several national and international advisory and grant awarding bodies in the areas of cybersecurity, IoT, and blockchain. He has been awarded more than £1.5M as Principal/Co-Investigator for 13 research projects. He is a cofounder of the CupCarbon IoT and smart cities simulator.

1 Hyperspectral Imagery Applications for Precision Agriculture
A Systemic Survey

Chanki Pandey, Yogesh Kumar Sahu, Prabira Kumar Sethy, and Santi Kumari Behera

CONTENTS

1.1 INTRODUCTION

Due to the current world's population, the demand for food will continue to rise. Unfortunately, environmental waste has reduced the percentage of arable land. As a result, farming and livestock yield activities are gaining prominence to satisfy the overflowing food demand. Precision agriculture is a method for optimizing productivity to fulfill increasing food requirements while minimizing economic and environmental costs related to food production. Precision farming includes extensive knowledge of temporal and spatial changes in crop conditions collected by remote sensing (Gevaert et al. 2015). The development of technologically advanced instruments for agricultural applications is important in helping farmers make crop health and resource management decisions. In addition to helping farmers efficiently use herbicides and pesticides, successful hyperspectral imaging in precision agriculture also offers insights into the current phase of crop growth and health. Precision

DOI: 10.1201/9781003152392-1

agriculture uses cutting-edge technology to make farming more regulated and precise, and imaging technologies, especially multispectral and hyperspectral imaging, are gaining prominence worldwide. According to the most recent Persistence Market Research (Persistence Market Research 2016) report, the demand for global imagery technologies in the precise agriculture market is expected to grow at a 9.0% CAGR from 2016 to 2024 at a market revenue of $1,165.9 million. According to the researchers, the precise agriculture imaging technology market produced $567.4 million in 2016, a 6.2% growth over 2015.

In large application areas such as mineral identification, environmental analysis, precise agriculture, and urban planning, the hyperspectral images can differentiate between different artifacts and physicals. Among other end-use sectors, forestry and agriculture are expected to have the largest market share in the hyperspectral imaging markets (Fact.MR 2019), as shown in Figure 1.1. In forestry and agriculture, hyperspectral imaging is used for various applications such as plant recognition, seed output analysis, weed mapping, and forest management. Furthermore, the amount of data gathered on farms from sensors has risen significantly over the last decade. Hyperspectral service providers have synchronized offers for data optimization and applications for farmers.

Nutrient crop surveillance, water stress, disease, insect infarction, and overall plant health are key components of effective farming operations. "The assurance of appropriate spatial and spectral knowledge for the non-destruction assessment of food and agricultural products is limited by traditional optical sensing technologies such as imaging or spectroscopy" (X. Li et al. 2018). In general, traditional imagery cannot obtain spectral information, and measurement by spectroscopy cannot cover vast areas of the sample. The commonly used fruit and vegetable sorting vision systems are based on a color video camera that emulates the vision of the human eye by taking photographs with red, green, and blue (RGB) wavelengths (Costa et al. 2011; Cubero et al. 2011). As a result, they are restricted to analyzing scenes. They are normally unable to detail the exterior or internal structure of the materials or spot flaws

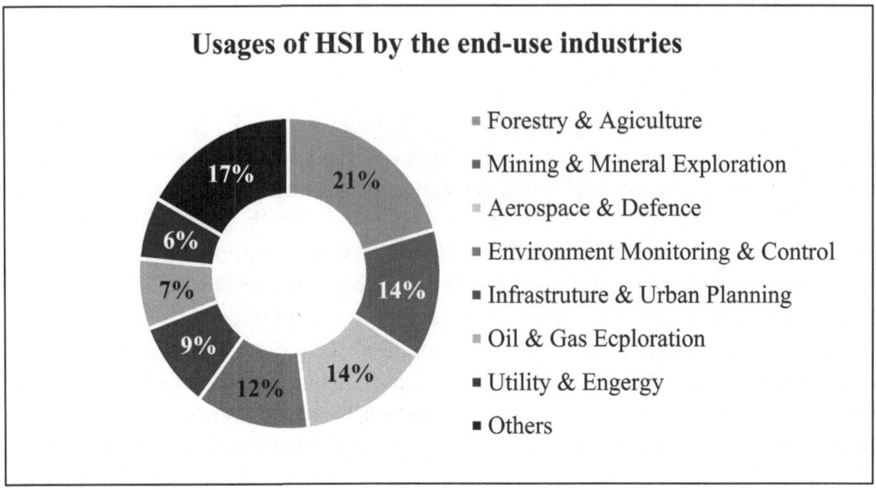

FIGURE 1.1 Usages of hyperspectral imaging (HSI) by the end-use industries.

or changes whose color is close to the color of the tough skin. Furthermore, conventional methods of tracking fruits and vegetables requiring analytical techniques are too time-consuming and costly and necessitate sample destruction.

Over the last few decades, optical sensor systems have evolved as scientific instruments for precision agriculture, thanks to the exponential advancement of information science and the analysis of images and patterns. In particular, spectroscopy and imaging techniques have been extensively researched and developed by incorporating a method that can obtain a spectral variance spatial map, which results in many popular applications in agriculture. The advancement of aerial and ground-based hyperspectral and multispectral imaging technology has been a significant step in the extension and practical implementation of precision agriculture techniques. In addition to its predictive capabilities, this technology has enabled evaluating crop stresses, characterization of soils and vegetative cover, bruise detection in fruits and vegetables (Z. Du et al. 2020), moisture content estimation (Z. Wang et al. 2020), and yield estimation (Y. R. Chen et al. 2002). Some of the advantages of hyperspectral and multispectral imaging are low costs (compared with conventional scouting); reliable performance; easy to use, quick, nondestructive, and highly precise evaluations; and applications of wide variety. A conventional spectral image comprises a series of monochromatic pictures that correspond to different wavelengths. Hyperspectral image systems have a natural benefit over conventional machine vision (Y. R. Chen et al. 2002) and even human vision. Any appearance characteristics difficult or hard to extract with conventional computer vision systems can be extracted using hyperspectral imaging systems. The quality and safety evaluation of various agriculture and food products has proved to be a vast application for hyperspectral imaging, for example, lamb (Fowler et al. 2015; Kamruzzaman et al. 2011; Pu et al. 2014), fruit (S. Chen et al. 2015; Nogales-Bueno, Rodríguez-Pulido et al. 2015), fish (Kimiya et al. 2013; Menesatti et al. 2010; Wu and Sun 2013; H. Zhang et al. 2020), cereals (Bianchini et al. 2021; Caporaso et al. 2018; Feng et al. 2019; Fox and Manley 2014; Manley et al. 2009; Paliwal et al. 2018; Qiu et al. 2018; Rabanera et al. 2021; Rathna Priya and Manickavasagan 2021; Sendin et al. 2019, 2018; Wakholi et al. 2018; Zeng et al. 2019), milk (Forchetti and Poppi 2017; Jawaid et al. 2013; Lim et al. 2016), pork (D. Barbin et al. 2012a, 2012b, 2013; Jiang et al. 2021; Silva et al. 2020; Xie et al. 2015), beef (Dixit et al. 2020; ElMasry et al. 2013; Naganathan et al. 2008), carcass of poultry (Falkovskaya and Gowen 2020; Nakariyakul and Casasent 2009; Nyalala et al. 2021; Park et al. 2007), veggies (Gowen et al. 2008; Ji et al. 2019; Nguyen Do Trong et al. 2011; Nogales-Bueno, Baca-Bocanegra et al. 2015), and so on.

1.1.1 The Main Contribution of the Chapter

1. The basic concepts of hyperspectral imaging technologies are covered.
2. The basics of hyperspectral imaging for food safety inspection are discussed.
3. The benefits and drawbacks of hyperspectral imaging technology are explored.
4. A review of hyperspectral image- processing techniques is given.
5. The current chapter highlights and outlines many important topics in key research concerning hyperspectral agricultural imagery applications.

In this brief study, we examined the past of hyperspectral imaging, imaging systems, precision agriculture applications, and hyperspectral data-processing techniques. To collect precise agriculture data, researchers can use hyperspectral imaging systems. Given the basic literature understanding, the topics of hyperspectral image processing are briefly explored in agriculture. The review covers both fundamental analyses of research and somewhere within the social facets of research. This analysis mainly seeks to identify potential works addressing the key issues and speed up deployable end-user solutions to achieve existing global goals for sustainable development. Section 2 contains simple hyperspectral image information and tools for interpretation. Section 3 analyzes the benefits and disadvantages of the various approaches used for hyperspectral imaging in precision agriculture. Section 4 provides a concise summary and concludes this chapter with the future scope of the hyperspectral imagery system (HSI) for precision agriculture.

1.2 HYPERSPECTRAL IMAGING TECHNOLOGY

A technology that combines traditional imaging and spectroscopy to obtain spatial and spectral information from an object simultaneously is developing, known as hyperspectral imaging or chemical and spectroscopic imaging. Goetz et al. (1985) first described the term "hyperspectral imaging," which was used to identify the surface material in the form of images directly. While the hyperspectral imaging technology was initially designed for remote sensing, the natural advantages over conventional machine vision (B. Zhang et al. 2014) are increasingly being demonstrated in a wide range of fields, such as agriculture. Hyperspectral imagery has recently developed into an effective technical inspection and consistency measurement method for fruits and vegetables by developing optical sensing and imaging techniques. The objective of hyperspectral imagery is to acquire the spectral range in the image of a scene for each pixel for target, substance identification, or process detection (Chang 2003). Hyperspectral picture is usually combined with spectroscopic methods, two-dimensional geometric space, and one-dimensional spectral detail detection to provide high spectral resolution and small band image results. Hyperspectral imaging (also referred to as imaging spectroscopy or imaging spectrometry) has been extensively investigated and developed, leading to many effective applications in the quality measurement of the precision agriculture by combining both the spectroscopic and imaging techniques into a single device that can obtain a spatial map of spectral variation. Table 1.1, however, displays the key variations between imaging methods, spectroscopy, and hyperspectral imagery.

In precision agriculture, hyperspectral imaging has practical uses, where the health condition of crops can be evaluated based on their specific signatures in various growth stages. Hyperspectral image sensors are space-sensing instruments capturing a scene's spectral behavior in the form of several simultaneous digital images, each reflecting an enclosed or nonstop spectral spectrum. As a certain substance is exposed to a light source with a known spectral bandwidth, it emits, absorbs, and/or reflects particular portions according to its structure. This reaction is called the spectral signature of a material (Manolakis et al. 2016). As shown in Figure 1.2, this information is contained in a cubic data structure, in which each spectral band

TABLE 1.1
The Key Variations between Imaging Methods, Spectroscopy, and Hyperspectral Imagery

Factors	Imaging Methods	Spectroscopy	Multispectral Imaging	Hyperspectral Imaging
Spatial information	√	×	√	√
Spectral information	×	√	Limited	√
Multiconstituent data	×	√	Limited	√
Detection of small-sized items	√	×	√	√
Spectral extraction's adaptability	×	×	√	√
Generation of quality-attribute distribution	×	×	Limited	√

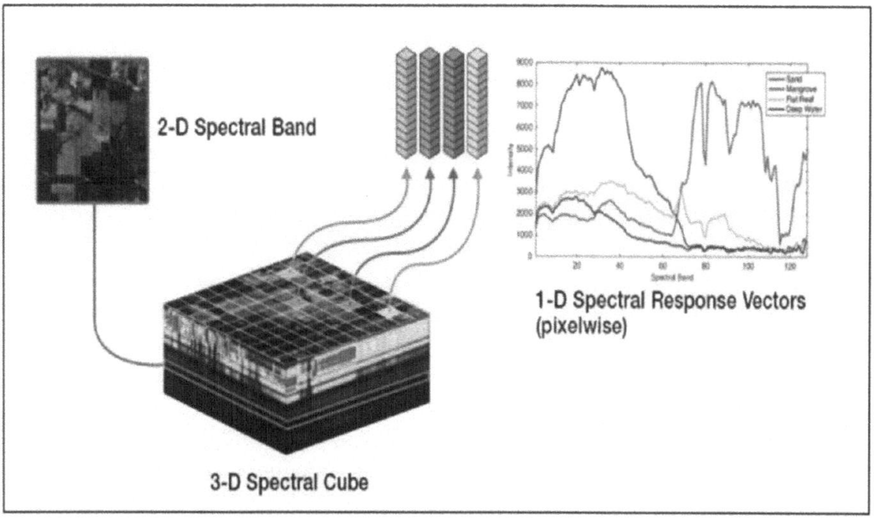

FIGURE 1.2 Hyperspectral Image signal model, showing a three-dimensional hyperspectral cube as a "stack" of individual spectral bands and individual hyperspectral pixel spectral responses.

Source: Arias et al. (2020)

is "stacked" by its wavelengths. The observations of measured spectral responses permit the classification of various materials or the observation of particular compositional qualities in biological subjects to be achieved through this information. Although the hyperspectral image's data volume is always extremely high and suffers from colinearity issues, the extraction of essential intimate analysis requires chemometric algorithms. A flowchart shown in Figure 1.3 illustrates typical stages of a complete algorithm to analyze hyperspectral images. Environment for Visualizing Images (ENVI) applications, MATLAB, and Unscrambler are the most commonly used commercially distributed software tools for hyperspectral imaging.

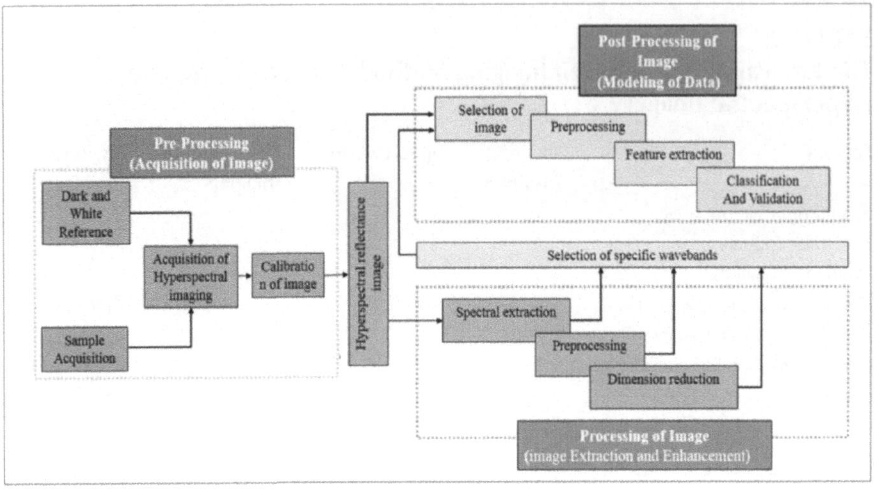

FIGURE 1.3 The typical stages of a complete algorithm to analyze hyperspectral imaging.

This technology has been used to control water and flood management to measure nutritional safety, medical diagnoses, military uses, and many more (Fei 2020; Khan et al. 2018; J. Ma et al. 2019). The advantages of hyperspectral image analysis in precision agriculture, which are particularly significant to the present work, are vital for monitoring the impact on the spectral response of plant tissues such as fertilizer (Serranti et al. 2018; Y. Wang et al. 2018; Zheng et al. 2018), micronutrient levels (Abdel-Rahman et al. 2017; Femenias et al. 2021; N. Hu et al. 2021; Jiang et al. 2021; W. Li et al. 2021; Mahajan et al. 2017; Pandey et al. 2017; Tahmasbian et al. 2021), harmful pests (Bock et al. 2010; Lowe et al. 2017; Moghadam et al. 2017; Rady et al. 2017; Susič et al. 2018), pollutant intakes (B. Huang et al. 2018; Lassalle et al. 2021; Stuart et al. 2019), or extreme conditions such as droughts and floods in localized areas. Further processing of spectral data can assess the impact of these variables on crop protection and productivity by calculating differences in spectral responses. For high spectral resolution images, the design of hyperspectral image systems involves complicated optomechanical components, which, because of the increasing weight, volume, and power requirements, limits the applicability of them. The instruments may be placed on different platforms such as satellites (Pearlman et al. 2003; Y. C. Tian et al. 2011), aircraft (Green et al. 1998; Mozgeris et al. 2018), UAV, and other portable appliances (Ishida et al. 2018; Vanegas et al. 2018) that are available for field and laboratory use. Since each platform offers various trade-offs in terms of spatial resolution, spectral resolution, measurement noise, range, and implementation costs, selecting the best imaging platform is application-specific.

1.3 AGRICULTURAL APPLICATIONS

This section illustrates and outlines the most significant contributions to different quantities of importance to hyperspectral imaging crops of rice. Figure 1.4 illustrates

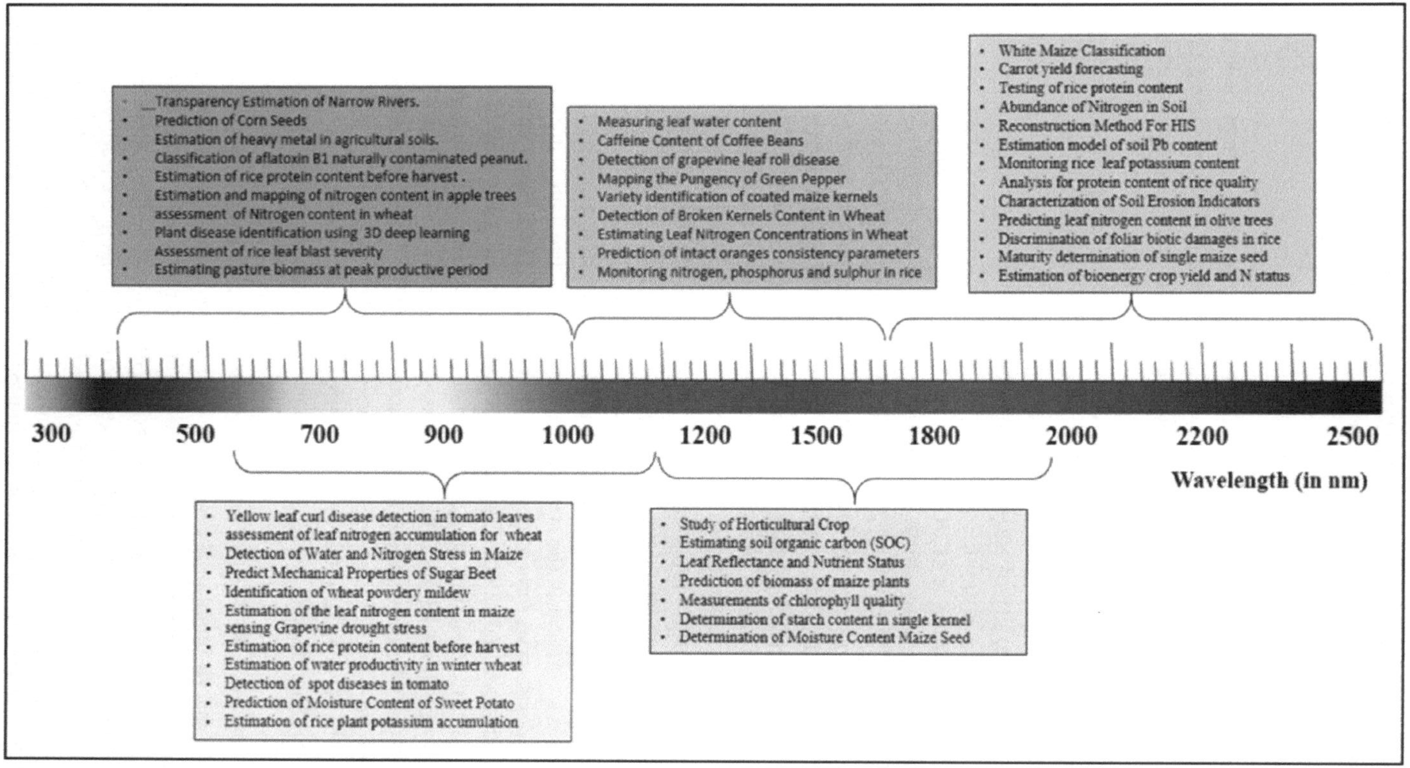

FIGURE 1.4 The application of hyperspectral imaging to various agriculture sectors based on the spectral range in nanometers.

the application of hyperspectral imaging to various agriculture sectors based on the spectral range in nanometers. The following subsection describes and analyzes the different specific applications of hyperspectral imaging in precision agriculture.

1.3.1 Soil Analysis

Development and land utilization practices have a long history within the Mediterranean locale misusing soils as a shared resource. The soils are a primary figure contributing to the rural generation, and their poor management is endangering their quality and efficiency. The climatic conditions of regions heavily determine the many factors of soil and soil erosion conditions like intense sun and heavy rain, affecting soil erosion. This further affects the quality and quantity of crops grown, so we need to determine some soil erosion aspects to counter agricultural problems. For estimating a variety of soil characteristics, the following factors are important for agriculture, such as salinity (J. Hu et al. 2019), fertility (Patel and Ghosh 2019), heavy metal (Pallottino et al. 2018; Shen et al. 2019; Shi et al. 2016; S. Zhang et al. 2019), pH (Grafton et al. 2019), and carbon (Bangelesa et al. 2020; Y. Chen et al. 2019; L. Guo et al. 2019, 2021), and hyperspectral images have also been used for this. The amount of nutrients found in the soil is an important factor to consider in precision agriculture. The availability of nutrients in the soils is essential to agricultural productivity, food safety, and sustainable agroecological growth (Y.-Q. Song et al. 2018). The soil properties are general, and in forecasting models should be taken into account for the reduction of soil nutrient loss and for improving soil fertilization managerial practices, based on more rigors and more efficient analysis techniques, to complement plant details such as soils' moisture, organic carbon, organic matter, and soil nutrient levels. Soil organic carbon (SOC) in topsoil can be calculated to help increase soil quality and food production. Since soil organic matter (SOM) is so critical in the soil–plant environment, researchers are searching for easier, nondestructive ways to detect it. In Hong et al. (2020), based on an RF model and ground-measured samples, an airborne hyperspectral image's ability to retrieve the SOC of bare topsoil in an agricultural site was investigated. The studies utilize this facility in about 1 km of the landmass in agricultural site southeast part of the United States, where the annual mean temperature is 9.8°C and land elevation range is 237–266 m.

Similarly, Reis et al. (2021) analysis aims to see whether HSI in the lab could forecast organic matter using multivariate regression modeling procedures. In Paraná, Brazil, 384 soil samples were obtained at eight depths in an Oxisol to group the sets according to depth. The spectrum was subjected to principal component and linear discriminant analysis (LDA). The photographs were categorized based on the quality of organic matter using HSI and the regression coefficient from the model. The partial least squares regression (PLSR) model was generated to compare sensor spectral data with SOM contents collected using the traditional method. The results obtained in the prediction were $R^2 = 0.75$, $r = 0.87$, and RPD = 2.1. In 2021, a proposed methodology supporting a hyperspectral imaging technique using LA-ICP-MS shows the spatial distribution of elements in soil cores (Zaeem et al. 2021). The LA-ICP-MS is a robust analytical technique that can provide quality representations of the spatial distribution of the components in various samples. Laser ablation is combined inductively with the plasma mass spectrometry, and the proposed method is capable of analyzing soil cores, including the levels and spatial distribution of Ca, Mg, P, K, Na, Zn, Fe, Co, and Mn. Moreover, previous studies have shown the potential for the identification of fungus detection methodologies and toxins in crops and food products, which impair plants' productivity (Gold et al. 2020; Gorretta et al. 2019; Mahlein et al. 2019; Yan and Yu 2019), lead to additional food waste levels (Parfitt et al. 2010), and cause adverse human health consequences during consumption (S. Liu et al. 2017; J. Wang et al. 2020; Xing et al. 2019). Table 1.2 summarizes the most relevant works in hyperspectral imaging applications for soil analysis.

TABLE 1.2
Summary of Most Relevant Work on Hyperspectral Imaging for Agricultural Applications for Soil Analysis

Objective	Reference	Method/ Devices Used	Spectral Range (in nm)	Result	Pros/Cons
Characterization of soil erosion indicators	(Schmid et al. 2016)	MEDSES, SEAS, SVM classifier	1,000–2,500 nm	Producer's maximum accuracy = 84% User/s maximum accuracy = 90% with kappa(κ) = 0.74	Large landmass data used; however, accuracy can be desired is less
The abundance of nitrogen in soil	(Patel et al. 2020)	DASU DASU-based DL network	350–2,500 nm 700–1,500 nm 2,100–2,500 nm	46.6% N for g/100g sample	Dataset used is remarkable
Estimating soil organic carbon (SOC)	(Hong et al. 2020)	Fractional-order derivative (FOD), random forest (RF)	380–1,000 nm, 900–1,700 nm	highest R_{CV}^2 = 0.66	In the imaging spectral band 400 and 1,000 nm algorithm meats the result
Heavy metal estimation of agricultural soils	(Tan et al. 2020)	RF, SRF, RRF, GRF, HySpex VNIR-1600, and HySpex SWIR-384 sensor	400–1,000 nm, 1,000–2,500 nm	$R_p^2 = 0.75$, $R_p^2 = 0.68$, $R_p^2 = 0.74$ $RMSE_p = 5.62$, $RMSE_p = 8.24$, and $RMSE_p = 2.81$ (mg/kg)	Simple model
Detection of soil organic matter	(Reis et al. 2021)	Partial least squares regression (PLSR)		$R^2 = 0.75$, $r = 0.87$, RPD = 2.1	Highly reliable

(Continued)

TABLE 1.2 (Contiued)

Objective	Reference	Method/ Devices Used	Spectral Range (in nm)	Result	Pros/Cons
Analysis of the spatial distribution of soil cores' components	(Zaeem et al. 2021)	Laser ablation inductively coupled plasma mass spectrometry (LA-ICP-MS)	213 nm	This methodology illustrates how various land management schemes modulate multielements' spatial distribution in the soil	Massive implications for the conservation of soil ecosystems
Estimation model of soil Pb content	(S. Tian et al. 2020)	MLR, PCR	500–2,500 nm	R^2 = 0.724%, RMSE = 24.92%, and MRE = 28.22%	MLR is superior than PCR

1.3.2 Contaminants and Nutrient Estimation

The estimation of the nutrient and biomass in crops helps classify the crop conditions and the different categorizations of soil-characterized crops to support agriculture development for farmers or others. Rapid determination of the nutritional content of lettuce cultivars grown hydroponically (Eshkabilov et al. 2021) proposed a model where in-line measurements were conducted using hyperspectral imaging processing techniques for four lettuce varieties (Rex, Tacitus, Black Seed Simpson, Flandria). Planting plants in Rockwool cubes was done for 3 weeks with 0, 50, 100, 150, 200, 250, and 300 ppm nitrogen hydroponic nutrient solution. The hyperspectral images of the newly cut leaf blades were recorded in a wavelength of 400–1,000 nm with a hyperspectral sensor. For the classification of aflatoxin B1 (AFB1) naturally contaminated peanut, He et al. (2021) proposed a method that discriminates the number of peanuts infected between usual and natural AFB1. In the spectral range of 400–1,000 nm, two variations of peanut were imaged by a hyperspectral imaging system with the support of enzyme-linked immunosorbent assay (ELISA) testing.

The leaf nitrogen content (LNC) is the most common of these physical magnitudes in the current state of the art because of their closer connection to rice crop yields. While most of these methods aim to approximate hyperspectral measurements in N levels directly, attempts have been made to refine their precision by integrating external information sources such as remote measurements by LIDAR (L. Du et al. 2016; Zheng et al. 2018). LNC is a measure of nutrients in plants collected from the visible, near-infrared, or shortwave infrared data by an indirect factor using various technologies. It is used as a predictor of crop development, which is linked to the crop sector. The approach (Z. Li et al. 2016) to estimate the nitrogen concentrations in winter wheat leaves was proposed in 2016, comparing spectral characteristics.

The predictive strength and effect of the sample size have been assessed. PLSR and SVR are key outcomes. It suggested that they would be higher than the other two, with the calibration set's coefficient of r^2 for 0.82 and 0.81 and the validity set for NRMSE of 5.48% and 5.94% for both methods. The reliability of the BPN prediction was improved as if the sample size had grown. It is still not recommended to use a BPN sample size smaller than 80. These findings show that the LNC assessments for winter wheat are preferred to PLSR and SVR, and BPN is recommended if the sample sizes are appropriate.

To track the nitrogen content of wheat canopy leaves, X. Song et al. (2016) had investigated the relationships between N material and wheat canopy spectral data in the principal solar plane (SPP) and perpendicular plane at various observation angles based on a 2-year experiment for winter wheat in Lethbridge (Canada), Zhengzhou (China), and Kaifeng (China) growing under different cultivation practices. Rubio-Delgado et al. (2021) have suggested the PLSR model for LNC prediction with the spectrum (350–2,500 nm) for the olive tree that combined two or three wavelengths along with the preprocessing techniques (smoothing, SM; regular natural variate, SNV). Assessment of leaf nitrogen accumulation (LNA) for a different wheat technique like PLSR, SVM, RF is used for chlorophyll absorption regions in the spectrum range of 550–750 nm, as suggested by J. Guo et al. (2021). The SVM regression is the most accurate model if chlorophyll absorption characteristic band values are normalized. Also it was suggested that RSI (NBDI743, NBDI703) could be used to estimate LNA using univariate linear. Again, a 3D winter wheat canopy model, i.e., the PLS-R model, was proposed based on the simulation techniques, combined with light propagation modeling to simulate the apparent reflectance of any visible leaf in the canopy for a given total reflectance. The prediction of LNC has been considered in the spectral range of 100–400 nm. Similarly, Wen et al. (2020) "proposed an Optimized Red-Edge Absorption Area (OREA) index to improve the prediction accuracy of Vertically Integrated (VI's) leaf N content within the spring- and summer-sown maize canopies."

Estimating and mapping nitrogen in apple trees at canopy and leaf stages, Ye et al. (2020) presented rapid and nondestructive approaches, including a hyperspectroscopic image collection system ImSpector V10 (400–1,000 nm). Vario EL cube calculated the nitrogen content of apple leaves. Raw reflectiveness and reflectance of the first derivative are applied to the leaf material of nitrogen. To infer nitrogen content from reflectance, PLSR and multiple linear regressions (MLR) were performed. Recently, in 2021, L. Huang et al. (2019) presented a method based on red-edge parameters and fractional differential to estimate the nitrogen level of an apple tree canopy with the support of hyperspectral reflectance, SVM, and RF approaches. The correlation coefficient with nitrogen content and spectral parameters such as the red-edge peak area (Sr()) obtained by fractional differential and logarithmic transformation processing can be achieved at a rate of 0.6 or higher.

A technique based on HSI segmenting with active learning on estimating the quality of chlorophyll in the crops is presented by satellite images (Nandibewoor and Hegadi 2019). In the first step, the authors used the Sparse SMLR model to learn the post-probability distributions of classes. Second, knowledge obtains from the first step of the Markov random field segment to estimate spatial and edge data dependencies, along with the minimum marker-based forest. Estimating rice protein content before harvest (Onoyama et al. 2018), ground-based, high-resolution hyperspectral imaging (1 mm to 1 mm per pixel) was used to estimate brown rice protein before harvesting. The author compares the estimated acuteness of estimated rice protein content modeling from five regions of interest (ROIs) medium reflection: total objective, dark area, canopy area (leaves, yellow leaves, and ears). The sample area of the target was 0.85 m per 0.85 m. Monitoring leaf potassium content using hyperspectral vegetation indices in rice leaves, Jingshan Lu et al. (2020) showed that the shortwave infrared (1,300–2,000 nm) region's spectral reflectance, R, was sensitive to K levels and was significantly associated with rice leaf K content. From 350 to 2,500 nm was determined for the published K vegetation indices in rice.

Recently, in 2021, [58] focused on separating non-negative matrix factorization (NMF) for hyperspectral reflectance from the ground and UAV platforms, mitigating impact on the water and soil context for estimating rice plant potassium accumulation with the support of the PLSR model in the spectral range 450–950 nm. C. Ma et al. (2021) had presented a protein content prediction model in rice (with husk) using a HSI. The proposed model utilizes the partially lower quadrant regression (PLSR), the partially component regression (PCR), and the lowest quadrant portion, which includes a collection of 87 rice species in China within wavelengths of 938–2,215 nm, and it is the first developed multispectral calibration models across the full range (LS-SVR). Table 1.3 summarizes the most relevant work on hyperspectral imaging for agricultural applications for contaminants and nutrient estimation.

1.3.3 Inland Water and Moisture Estimation

The main component of agricultural water supply in agricultural irrigated regions is water productivity (WP). WP is used to establish relationships between crop production and the involvement of water for the production, also expressed as crop production per unit volume of water. In 2016, proposed estimating WP in winter wheat and biomass data based on HIS with the AquaCrop model. Spectral variables and concurrent biomass, yield, and samples were acquired at the Xiaotangshan experimental site in Beijing, China, during the winter wheat growing seasons. Using the HIS, Y. Sun et al. (2017) presented a method for measuring moisture content and freezable water content during the drying process of purple-fleshed sweet potato slices. The Savitzky–Golay smoothing filter and multiplicative scatter correction (MSC) are being deployed to preprocess the raw spectra and PLSR to analyze the relationship between spectral data and quality attributes in visible and near-infrared 371–1,023 nm spectral range.

Consequently, accurate calculation of the quality of the leaf water on the canopy scale is more difficult than on the scale of the leaf as canopy calculations provide details on areas of bare soil, green and non-green plant components, and shading. A small number of studies (Kim et al. 2015; Pandey et al. 2017) use hyperspectral images to measure water absorption on the surface of individual leaves. As a result, there are very few details about how water is transported and how this information can be used to diagnose water stress. Murphy et al. (2019) suggested spectral indicators such as the moisture stress index (MSI), the normalized difference water index (NDWI), and the sensitivity of particular water absorptions at 970–1,775 nm to

TABLE 1.3
Summary of Most Relevant Work on Hyperspectral Imaging for Agricultural Applications for Contaminants and Nutrient Estimation

Objective	Reference	Method/Devices Used	Spectral Range (in nm)	Result	Pros/Cons
Study of horticultural crop Leaf reflectance and nutrient status	(Nguyen et al. 2020)	Night-based hyperspectral imaging	N = 700–709 nm, K = 780–787 nm, Mg = 817–821 nm	Classification accuracy 75% (bok choy) and 80% (spinach) in spectral region 700–830 nm	Hyperspectral remote sensing (HSRS) data acquisition is not susceptible to shadows and physical structures within greenhouses
Biomass prediction of maize crops	(D. Ma et al. 2020)	Kernel partial least squares (KPLS)	370–1,030 nm, 465–962 nm	$R^2 = 0.924$	This new nonlinear KPLS model could also improve the precision of estimates of other plant properties
Measurements of chlorophyll quality	(D. Gao et al. 2021)	ROI-ALR-RF-PLSR with ROI-OMR-RF	451, 552, and 648 nm	$R_v^2 = 0.52$ and RMSE=3.61	R_v^2 increased RMSE decreased
Testing of rice (with husk) protein content	(C. Ma et al. 2021)	PLSR, PCR, and LS-SVR	938–2,215 nm	RMSE = 0.52 $R^2 = 0.8011$	The distribution of rice protein can be understood according to the visualization picture
Determination of starch content in single kernel	(C. Liu et al. 2020)	ANN, PLSR, adaptive reweighted sampling (CARS)	930–2,500 nm	$R_p^2 = 0.96$ RMSEP = 0.98	CARS was used to select the effective wavelengths
Classification of aflatoxin B1 naturally contaminated peanut	(He et al. 2021)	PLS-DA and SVM LDA	400–1,000 nm	Classification accuracy = 94%	For chemical analysis, spectral, texture, and color characteristics were derived and incorporated
Predicting leaf nitrogen content (LNC) in olive trees	(Rubio-Delgado et al. 2021)	PLSR, SVM, LNC, SM	350–2,500 nm	$R^2=0.72$, $R^2_{cv}=0.71$, $R^2=0.64$, $R_p^2=0.63$	Models based on PLSR were more accurate than models based on VI
Monitoring nitrogen, phosphorus, and sulfur in hybrid rice	(Mahajan et al. 2017)	VIs (NDVI hyper and NDVI broadbands)	670–1,460 nm	$r = 0.81$, $\rho < 0.01$ for predicting N $r = 0.58$, $\rho < 0.01$ for S predicting S	Results indicated that released VIs could retrieve canopy N with greater precision but not P or S

(Continued)

TABLE 1.3 (Continued)

Objective	Reference	Method/Devices Used	Spectral Range (in nm)	Result	Pros/Cons
Estimating leaf nitrogen concentrations in wheat	(Z. Li et al. 2016)	PLSR, SVR, BPN, SMLR	350–1,050 nm, 1,450 and 1,940 nm	$R^2 = 0.78$, RMSE = 6.52%	SVR outperformed the other two approaches in this study
Estimation of rice protein content before harvest	(Onoyama et al. 2018)	RGB histogram and a GNDVI—NDVI image processing	400–1,000 nm	$R^2 = 0.76$ and $RMSE = 0.35\%$	It is not needed for the rice protein content before harvest to obtain an image of a field resolution greater than 0.85 €/pixel
Assessment and mapping of apple-tree nitrogen content at leaf and canopy levels	(Ye et al. 2020)	ImSpector V10, Vario EL cube, four key wavelengths	505, 560, 675, and 705 nm 400–1,000 nm	The first derivative reflection PLS and MLR models: $R^2 = 0.7745$ and 0.774 ($p < 0.001$)	The MLR model has shown its edge over the PLS model
Estimate the content of chlorophyll from the plants	(Nandibewoor and Hegadi 2019)	SMLR-PSO, Markov random field segments	–	Accuracy = 95.0%	Experimental results demonstrate that the suggested methodology outputs advanced processes for real-world hyperspectral datasets in precision and time of execution
Assessment of nitrogen content in wheat	(Al Makdessi et al. 2019)	PLS-R	400–100 nm	Average error = 0.83%DM	Approach tested on plants in a greenhouse and the field
Estimation of rice plant potassium accumulation	(Jingshan Lu et al. 2021)	Non-negative matrix factorization (NMF), PLSR	450–950 nm	R^2 (PLSR) = 76% RMSE = 16.93%, RE = 45.07%.	NMF may be extended to both land and UAV hyperspectral platforms
Estimates the biomass of pastures for the peak harvesting season	(Tong et al. 2019)	PLSR, SMLR	450–900 spectral range	$RMSE_{CV} = 37\%$, RMSE = 12.5%	Combined use of *in situ* HSVI
Assessment of leaf nitrogen accumulation for wheat	(J. Guo et al. 2021)	PLSR, SVM regression, RF, RSI (NBDI743, NBDI703), chlorophyll absorption regions	550–750 nm	SVM: $R^2 = 0.895$ RMSE =0.903 gm^{-2}, univariate linear regression: $R^2 = 0.806$ RMSE =1.231 gm^{-2}	The SVM regression was found to be the most accurate model if chlorophyll absorption characteristic band values normalized

Objective	Reference	Method/Devices Used	Spectral Range (in nm)	Result	Pros/Cons
Estimation of the vertically integrated LNC in maize	(Wen et al. 2020)	Optimized red-edge absorption area (OREA)	680–760 nm	r^2=0.811, RMSE=0.374, RE=13.17%	The OREA index can be used to calculate the amount of N_2 in various field experiments
Monitoring rice leaf potassium content using HS-VI	(Jingshan Lu et al. 2020)	Normalized difference spectral index (NDSI)	350–2,500 nm	R^2= 0.68	The spectral range is high
Caffeine content of coffee beans	(C. Zhang et al. 2017)	PLSR, RF	874–1,734 nm	PLSR R_p^2= 0.843, RMSEP =131.904 µg/g, RF R_p^2= 0.878 and RMSEP = 116.327 µg/g	The spectral range is moderate
Estimation of nitrogen content on apple tree canopy	(Peng et al. 2021)	SVM, RF, Grunwald—Letnikov fractional difference algorithm	RF wavelength	$R^2_{SVM} R_p^2$ = 0.56, $RMSE_{SVM=}$ 1.51, R^2_{RF} = 0.94, $RMSE_{RF}$ = 0.84	RF model has a better predictive effect
Prediction of copper contamination in the wheat canopy	(G. Wang et al. 2020)	NDVI/SIPI and W728, the spectral resolution of 2.4 nm	350–1,000 nm	R^2 = 0.669 for NDVI/SIPI R^2 = 0.625 for W480	The model is simpler and covers a wide spectrum
Analysis for protein content of rice quality	(D. Sun et al. 2019)	NDSI, GWAS, phenotyping	350–2,500 nm	R^2_{NDSI}=0.68	It offered a new way to phenotype one of the main biochemical features to measure rice consistency for genetic studies
Detection of nitrogen stress in maize	(Naik et al. 2020)	IW/CPE ratio of 0.6, 0.8, and 1.2 nitrogen level treatments (300 kg of N ha^{-1})	540–860 nm	Leaf water content = 80%. Leaf area index = 61% and leaf nitrogen contentment = 66%	System accuracy is less
Detection of hydroponic lettuce cultivars' nutrient content	(Eshkabilov et al. 2021)	PLSR and PCA	506–601 nm and 634–701 nm	R^2 = 0.784–0.987	The amount of nutrients in lettuce is easily detected in hyperspectral imaging techniques
Estimation of bioenergy crop yield and N status	(Foster et al. 2017)	ASD FieldSpec FR spectroradiometer, NDVI, PLSR	350–2,500 nm	$RMSE_{PLSR}$ = 41 %. r^2 = 0.72 and RMSE = 4.6 Mg ha^{-1}	The RMSE of the model PLSR was 19%–41% lower than the best N-VI and was at a rate of 4.2%–100% lower for final biomass than the best narrowband VI

collect data in the lab. The midrib, the green areas of the leaves, and the entire leaves were all measured separately for absorption."Z. Wang et al. (2020) demonstrated a method for evaluating the MC (moisture content) of maize seeds, the long-wave NIR hyperspectral imaging (930–2548 nm) coupled with the UVE-SPA hybrid vector selection algorithm and LS-SVM model was successfully used. Table 1.4 summarizes the most relevant work on hyperspectral imaging for agricultural applications for inland water and moisture estimation.

1.3.4 Crop Yield Estimation

One of the most important research fields of precision agriculture is yield prediction. To maximize production, yield prediction, yield mapping, crop supply matching with demand, and crop management are crucial (Al-Gaadi et al. 2016). The estimate of crop yield is one of the key agricultural management problems and can provide the greatest value for precision farming techniques. For carrot yield forecasting, Suarez et al. (2020) presented a model based on proximal hyperspectral and satellite multispectral sensors (Sentinel-2 and WorldView-3), which examined the accuracy and prediction of carotid roots in three rising regions with distinct cultivation configurations, periods, and soil conditions. In 24 fields in Western Australia (WA), Queensland (Qld), and Tasmania (Tas), Australia, above ground biomass (AGB), canopy reflectances, and the related yield indicators have been obtained in 414 sampled areas. Forage yield and quality estimationare were being carried out by Geipel et al. (2021) in the range of 450–800 nm, and a UAV and an HSI were used to catch canopy reflections grass–legume mixture in Southeast and Central Norway. In the proposed method, the PPLSR, neutral detergent fiber (NDF), and SLR are used to determine the fresh (FM) and dry matter (DM) yields, as well as crude protein (CP) with $RMSE_{\text{Fresh matter}} = 14.2\%$, $RMSE_{\text{Dry matter}} = 15.2\%$, and $RMSE_{\text{crude protein}} = 11.7\%$, none of the SLR models that were tested had acceptable prediction accuracies. A significant goal in precision agriculture is an estimate of the fruit yield in orchard blocks, as it facilitates future planning and productive use of resources by farmers. Gutiérrez et al. (2019) introduced a new HSI field-based line-scan mango yield estimation pipeline obtained from an unmanned ground vehicle. Models on hundreds of trees were validated, checked, and mapped. For research, the coefficients of determinations were up to 0.75 for the field counts (18 trees) and 0.83 for the RGB mango counts (predicting 216 trees).

Similarly, the estimation of corn yield was being carried out by W. Yang et al. (2021), and the hyperspectral imaging technique was used to develop a CNN classification model. Five maize development stages captured high-resolution hyperspectral photographs—V5 (5 leaves in blue collars), V8 (8 leaves in blue-colored columns), V10 (10 leaves in blue-colored columns), V12 (12 leaves in visible blue columns), and R2 (blister stage). The wavelet analysis was used to denote hyperspectral imagery and then to train and test the CNN model. Fast and precise biomass estimate and yield estimate enable effective plant phenotyping and crop management on site. B. Li et al. (2020) proposed strategic methodology to estimate the above growth bio-mass and estimate crop yield, a low-altitude, unmanaged aerial (UAV), and hyperspectral imagery data was obtained for a potato crop canopy during two development steps.

TABLE 1.4
Summary of Most Relevant Work on Hyperspectral Imaging for Agricultural Applications for Inland Water and Moisture Estimation

Objective	Reference	Method/Devices Used	Spectral Range (in nm)	Result	Pros/Cons
Reconstruction of HIS from multispectral data for complex coastal and inland waters	(Banerjee and Shanmugam 2021)	MSI and HICO, DNN	438–868 nm	Successful reconstruction and validation of the hyperspectral radiances	Local, regional, and global applications of society
Measuring leaf water content	(Murphy et al. 2019)	NDWI, MSI	970–1,775 nm	$R^2 = 0.35$ NDWI = 0.83 MSI = 0.72	The effects of the quantification of hyperspectral imagery leaf water quality gained at space resolutions are high enough for individual components to be resolved
Estimation of water productivity in winter wheat	(Jin et al. 2018)	AquaCrop model	600–1,000 nm	$R^2 = 0.79$ and RMSE = 0.12 kg/m^3	5-year data collected and analyzed from 2008 to 2012
The moisture content of sweet potato slices of purple-fleshed	(Y. Sun et al. 2017)	Contact ultrasound-assisted hot drying (CUHAD), moving average, Savitzky–Golay smoothing filter, MSC	371–1,023 nm	$R_p^2 = 0.9323$, considered determining moisture content	A fast and nondestructive process for estimating moisture content and freezable water content at various dehydration times
Detection of water stress in maize	(Naik et al. 2020)	IW/CPE ratios of 0.6, 0.8, and 1.2, nitrogen level treatments (300 kg of N ha^{-1})	540–860 nm	Leaf water content = 80% Leaf area index = 61% and leaf nitrogen contentment = 66%	System accuracy is less
Moisture content of a single maize seed determination	(Z. Wang et al. 2020)	LWNIR, HIS, UVE-SPA-LS-SVM, PLSR	930–2,548 nm	Correlation coefficient of prediction (R_{pre}) = 0.9325 RMSEP = 1.2109	A wide range of seed samples of maize would still further refine the proposed process
The moisture content of a single maize seed is estimated.	(Z. Wang, Fan et al. 2021)	Competitive Adaptive Reweighted Sampling (CARS) and Successive Projections Algorithm (SPA)	930–2,548 nm	Prediction accuracy CARS-SPA-LS-SVM $R_{pre} = 0.9311 \pm 0.0094$, RMSEP = 1.2131 ± 0.0702 in 200 independent assessment	The highest prediction results are in CARS-SPA-LS-SVM with mixed spectra.
Transparency estimation of narrow rivers	(Wei et al. 2020)	XGBoost, spatial resolution is 18.5 cm GBDT, Ada Boost	400–1,000 nm 400–430 nm 680–840 nm	XGBoost regression accuracy = 97% RMSE = 2.515 cm MAE=2.008 cm	1. RMSE is lowest, and MAE is moderate 2. Simpler calculation-based algorithm

Six cultivars and various nitrogen treatments, potassium, and mixed compound fertilizers were used as field tests. Crop height was measured using the disparity between the optical surface model and the RGB imaging models combining with the RReliefF function collection algorithm of two narrow-band vegetation indexes. Models for random forest regression showed good predictive precision with a determination factor (r^2) > 0.90 for both fresh and dried above-ground biomass. Four narrow-band vegetation and crop height indices have been projected to yield crops, and images are collected 90 days after plantations ($r^2 = 0.63$). An improvement in the yield prevision was shown by a partial regression model based on the maximum wavelength spectrum ($r^2 = 0.81$). Table 1.5 summarizes the most relevant work on hyperspectral imaging for agricultural applications for inland water and moisture estimation.

TABLE 1.5
Summary of Most Relevant Work on Hyperspectral Imaging for Agricultural Applications for Crop Yield Estimation

Objective	Reference	Method/Devices Used	Spectral Range (in nm)	Result	Pros/Cons
Estimation of corn yield	(W. Yang et al. 2021)	CNN + HSI	–	Kappa coefficient $\kappa = 0.69$ Classification accuracy = 75.50%	Estimates of corn yield using CNN offered insights on yield improvement
Carrot yield forecasting	(Suarez et al. 2020)	Satellite multispectral sensors (Sentinel-2 and WorldView-3)	350–2,500 nm	$R^2 < 0.57, p < 0.05$	The approach gives considerable value to logistics management
Estimation of the yield and N status of bioenergy crops	(Foster et al. 2017)	ASD FieldSpec FR spectroradiometer, NNVI, and PLSR	350–2,500 nm	$RMSE_{PLSR} = 41$ %. $r^2 = 0.72$, and RMSE = 4.6 Mg ha^{-1}	The findings show that PLSR could be the best predictor of final biomass yield using spectral
Forage yield and quality estimation 2021	(Geipel et al. 2021)	UAV, PPLSR, neutral detergent fiber (NDF), SLR	450–800 nm	$RMSE_{Fresh\ matter} =$ 14.2% $RMSE_{Dry\ matter} =$ 15.2% $RMSE_{Protein\ content} =$ 11.7%	None of the SLR models that were tested had acceptable prediction accuracies
Prediction of corn seeds	(Pang et al. 2020)	CNN(1D and 2D)+ HSI	400–1,000 nm 900–1,710 nm	Recognition accuracy = 90.11% (1DCNN) Recognition accuracy = 99.96% (2DCNN)	Fast convergence speed

1.3.5 Plant Disease Monitoring, Insect Pesticide Monitoring, and Invasive Plant Species

Farmers face major threats due to the emergence of various pests and diseases in the crops. Fungi, bacteria, viruses, and nematodes are some of the common reasons for disease infections. Traditionally, most of the diseases were not diagnosed or suspected by the farmers as they lack knowledge about the crop diseases and solid required support and suggestions from the specialist. Hence, the diagnosis of disease infections in their early stage can protect the crops from real damage. For identifying plant diseases, image processing plays an important role in detecting and identifying diseases only done by visual information. Zovko et al. (2019) proposed a model based on PLS-DA that was used to interpret the images, and PLS-SVM was used to classify the therapies for sensing grapevine drought stress. This study was conducted in a test vineyard in Dalmatia, Croatia, grown in an artificially transformed karst terrain with a precision of more than 97%; PLS-SVM showed the potential to assess degrees of grapevine drought or irrigated treatments. In the past few decades, MSI and HSI cameras installed across aircraft and drones in visible and NIR zones have proved to be an efficient way of evaluating crop infection in a swift, reliable manner. This way, it is possible to significantly minimize the time required to diagnose potential crop infection and simulation by adjusting the number of pesticides used for particular field necessities by adopting selective pesticide tragedies. To highlight the internal quality of apple fruit slices, Lan et al. (2021) presented a novel method of assessing the heterogeneity of apple fruit that is emphasized in two successive stages with close-infrared hyperspectral imaging (NIR-HSI). The pictures of NIR-HSI were acquired on the cut surface of six cross-slices per apple, systematically tested with a slice of five to six cylinders. The NIR-HSI picture PCA permitted 141 representative cylinders from the overall dataset to be chosen (samples 1,056), with spectrophotometric and chromatographies quantified to include the contents of dry matter, total sugar (TSC), fructose, glucose, sucrose, malic acid, and polyphenols. The SVM model is proposed by Nagasubramanian et al. (2018) to identify charcoal rot disease in soybean stems with a spectral range of 383–1,032 nm and an average overall accuracy of 90.91%.

The most broadly utilized practice in irritation and disease control is to shower pesticides over the cropping area consistently. This practice, albeit powerful, has a high financial and significant ecological expense. Discrimination of foliar biotic damages in rice was carried out by Z. Y. Liu et al. (2018), with the help of the hyperspectral reflectance of symptomatic and asymptomatic rice leaves diseases. Which are affected with Pyricularia grisea Sacc, Bipolaris oryzae Shoem, Aphelenchoides besseyi Christie, and Cnaphalocrocis medinalis Guen was measured in the 350–2,500 nm spectral range in a laboratory. For yellow leaf curl disease detection in tomato leaves, Jinzhu Lu et al. (2018) used the HSI technique with the spectral range of 500–1,000 nm. Both background and the leaf area were analyzed to select sensitive wavelengths and band ratios, using a grey level co-occurrence matrix (GLCM), 24 texture features for extraction. Similarly, for the detection of target spot (TS) and bacterial spot (BS) diseases in tomatoes, Abdulridha et al. (2020) proposed a methodology based on multilayer perceptron neural network (MLP), stepwise

TABLE 1.6
Summary of Most Relevant Work on Hyperspectral Imaging for Agricultural Applications for Plant Disease Monitoring, Insect Pesticide Monitoring, and Invasive Plant Species

Objective	Reference	Method/Devices Used	Spectral Range (in nm)	Result	Pros/Cons
Detection of grapevine leaf roll disease in a red-berried wine grape	(Z. Gao et al. 2020)	Monte-Carlo, SVM, GLD	690, 715, 731, 1,409, 1,425, and 1,582 nm	MAX classification accuracy = 89.93%	Six wavelengths (690, 715, 731, 1,409, 1,425, and 1,582 nm) were sensitive to virus symptoms
Yellow leaf curl disease detection in tomato leaves	(Jinzhu Lu et al. 2018)	Grey level co-occurrence matrix (GLCM)+ SVM	500–1,000 nm	Excellent results	A validation set's discrimination result was 100% based on its best threshold value
Detection of target spot (TS) and bacterial spot (BS) diseases in tomato	(Abdulridha et al. 2020)	Multilayer perceptron neural network, stepwise discriminant analysis	380–1,020 nm	Classification accuracy = 99% for both TS and BS	In four types, tomato leaves were classified: stable, asymptomatic, early, and late-stage development of the disease
Three-dimensional deep learning for plant disease recognition	(Nagasubramanian et al. 2019)	DCNN	400–1,000 nm	Classification accuracy = 95.73%, F1 score = 0.87	The approach focuses on an economically critical illness, red charcoal, a fungal disease that affects the worldwide production of soybean crops
Identification of soybean stem charcoal red disease	(Nagasubramanian et al. 2018)	SVM	383–1,032 nm	Overall accuracy 90.91%	Simplex system
Diagnosis of plant cold damage	(W. Yang et al. 2019)	GLPF, Savitzky–Golay smoothing, CNN	450–885 nm	Coefficient of correlation = 0.8219. Computational efficiencies: W22 = 41.8%, BxM = 35%, B73 = 25.6%, PH207 = 20%, Mo17 = 14%	High throughput

Remote-sensing grapevine drought stress	(Zovko et al. 2019)	PLS-DA and treatments were classified using PLS-SVM	409–988 nm	Accuracy = 97%	PLS-DA identified relevant wavelengths, which were linked to O—H, C—H, and N—H stretches in water
Rice foliar biotic damage classification	(Z. Y. Liu et al. 2018)	Principal component analysis and probabilistic neural network	350–2,500 nm	The overall accuracies for *R*, lg (1/*R*), R′ and (lg (1/*R*))′ were 97.7, 98.1, 100, and 100%, and the Kappa coefficients were 0.962, 0.97, 1, and 1, respectively	Under laboratory conditions, the findings demonstrated that hyperspectral remote sensing could distinguish between different diseases and insect damage to rice leaves
Identifying corn seeds with varying degrees of freezing damage	(J. Zhang et al. 2021a)	KNN, DCNN	450–979 nm	Accuracy = 97.55%	Complex but accurate
Assessment of rice leaf blast severity during late vegetative growth	(G. S. Zhang et al. 2020)	SVM, SRR	400–1,000 nm	Classification accuracy of jointing stage = 83.33%, booting stage = 97.06%, and heading stage = 83.87%	The SRR data reconstruction approach presented here can be used to evaluate the magnitude of the rice leaf blast during late vegetative development, as our results show
Broken kernel content of bulk wheat samples detection	(Ravikanth et al. 2016)	PCR, PLSR, ten-fold cross-validation, LDA, QDA	1,000–1,600 nm	MSEP = 0.483 SECV = 0.70 $r = 0.94$ Classification accuracy QDA = 89.8% ± 2.6%, LDA 87.7% ± 1.6 %	PLSR performed better than PCR

discriminant analysis with the spectral range of 380–1,020 nm, and initial disease symptoms caused by different pathogens were the early lesions of TS caused by the fungus *Corynespora cassiicola* and BS caused by *Xanthomonas perforans.* At the same time, the image is collected by the UAV. Table 1.6 summarizes the most relevant work on hyperspectral imaging for agricultural applications for plant disease monitoring, insect pesticide monitoring, and invasive plant species.

1.3.6 Agricultural Crop Classification

The identification and classification of the crop (CI) using HSIs is an important field for research that considers different applications associated with agriculture. By Tan et al. (2018), an adaptive learning method for the Powell selective variance and Q-N multinomial LR classification was proposed. The (logistic regression). Here, the solution had dual steps, and the first step concerned the swift method for the MLR classification. The second step concerned the selection of the most suited unlabeled samples. In addition, SVM is used to select the most informative unlabeled samples based on PDD (posterior density distribution). L. Wang et al. (2020) has proposed a compact and low-cost HSI handheld platform (called LeafSpec) for crop leaf imaging with even better measurement accuracy than conventional HSI systems. In the experiment, they used the push-boron camera for HIS and the encoder system for leaf positioning and claimed that implemented technology could be helpful between two nitrogen treatments of corn plants in each genotype. Crop freezing destruction is a major agricultural catastrophe that affects the quality assurance of seeds. J. Zhang et al. (2021a) examined the possibility of integrating hyperspectral imaging with a deep convolutionary neural network (DCNN) to classify various freeze-damaged maize plants. The findings indicate that the DCNN model has achieved the most satisfactory results with accuracy rates in the five classification of 100% (entraining, validity, and testing), 96.9% (validation set), and 97.5% (testing set) with accuracy rates of 100% for the four-category classifications as well as DCNN model performing best in the assessment indexes. The prediction and mapping of the capsaicin and dihydrocapsaicin contents in green pepper were carried out by Rahman et al. (2018) using hyperspectral imaging. The best result was obtained by normalizing the preprocessed spectra with correlation coefficients (R_{pred}) of 0.86 and 0.59, respectively, using PLSR with various spectral preprocessing techniques; the best performance was found by normalizing the preprocessed spectra with correlation coefficients (R_{pred}) of 0.86 and 0.59, respectively.

The prediction of plant physiological characteristics, including biomass, using hyperspectrum imaging systems for high-throughput plant phenotyping, was suggested using multivariable spectral data modeling approaches. However, most of the proposed models are linear models such as an approximate area of leaves or plant biomass regression model that does not reflect a properly nonlinear association among the hyperspectral image data and anticipated phenomena. da Conceição et al. (2021) developed a rapid system for identifying fungi (*Fusarium verticillioides* and *Fusarium graminearum*) using near-infrared hyperspectral images combined with pattern recognition analysis and partial least squares discriminant analysis (PLS-DA) of images. Fifteen *Fusarium* isolates were used to validate the HSI-NIR process and conduct subsequent research. For white maize classification, Sendin et al. (2019) presented a method based on grading regulations and stipulated in South African. In this method, the types of undesirable materials regarded were divided into 13 classes. Using principal component analysis (PCA) and partial least squares discriminant analysis (PLS-DA) modeling, two approaches to data analysis, pixel-wise and object-wise, were studied. This approach results in a high occurrence of errors with a classification accuracy of 63–99%.

The maize seed maturity classifications are very important since they may improve yields. Z. Wang, Tian et al. (2021) presented a maturity classification of maize seeds using near-infrared hyperspectral imaging (NIR-HSI). The hyperspectral photographs of the embryo and endosperm side of corn seeds were taken in the spectral range of 1,000–2,300 nm to detect the effect of spectra of different locations in maize seed for modeling. Overall, the study found that long-wave near-infrared hyperspectral imaging can be used to noninvasively and quickly quantify the MC in single maize seeds and that a robust and accurate model based on the CARS-SPA-LS-SVM system combined with mixed spectral imaging can be developed. These findings may be used to evaluate other internal consistency characteristics (such as starch content) of a single maize crop. Table 1.7 summarizes the most relevant work on hyperspectral imaging for agricultural applications for agricultural crop classification.

1.4 CONCLUSION AND FUTURE SCOPE

This review provides a comprehensive overview of recent applications and developments of HIS in agriculture. In this review article, the recent hyperspectral imaging technology application developments since 2015 are highlighted in evaluating the perceptual properties of various agricultural products, suggesting that hyperspectral imaging technology would have significant potential as a fast and noninvasive visual analysis tool. As most research relies on the study of spectral information and some studies have already demonstrated the effectiveness of the mixture of spectral and spatial information, we suggest that more studies on the use of data fusion techniques should also be performed. Multiband hyperspectral cameras could be combined with less costly devices such as smartphones, allowing for widespread use of this technology in everyday life. Most used spectral region (400–1,000 nm). Most applied multivariate analysis is linear regression models includes PLSR and MLR, CNN, SVM, Savitzky–Golay smoothing, and RF.

The current analysis also highlights the common challenges facing the hyperspectral rice farming systems, which we believe are highly significant in developing the widely deployable rice cultivation technology applications, aiming to support effective agro-industrial rice cultivation. To achieve global sustainability needs, the broad rollout of the surveyed technical developments is key, a significant challenge in the current era. If such a methodology is successfully integrated into the decision-making process, a much more precise import/export rate will dramatically minimize future food waste and economic losses. For these purposes, an emphasis on developing low-cost, user-friendly spectrum imaging instruments that can be deployed more efficiently for growing operations of varying sizes is very important in future research.

TABLE 1.7
Summary of Most Relevant Work on Hyperspectral Imaging for Agricultural Applications for Agricultural Crop Classification

Objective	Reference	Method/Devices Used	Spectral Range (in nm)	Result	Pros/Cons
White maize classification	(Sendin et al. 2019)	PLS-DA, PCA	1,118–2,425 nm 1,219 and 1,476 nm (associated with starch), 1,941 nm (associated with moisture), and 2,117 nm	Classification accuracy of PLS-DA = 63–99%	Simplex system
Corn seed variety classification	(J. Zhang et al. 2021b)	DCNN, SVM, KNN	450–979 nm	DCNN training accuracy = 100% Testing accuracy rate = 94.4%, and validation accuracy rate = 93.3%.	DCNN can classify corn seed varieties using spectral data processing and have better performance than the other two
Mapping the pungency of green pepper	(Rahman et al. 2018)	PLS, successive projections algorithm (SPA)-PLS	1,000–1,600 nm	R_{pred} = 0.86 SEP = of 0.09	Green pepper pungency can be assessed quickly
Emphasizing the intrinsic consistency of apple fruit slices	(Lan et al. 2021)	Near-infrared hyperspectral imaging (NIR-HSI)	–	R_{cv}^2=0.83, RPD=2.39 and TSC R_{cv}^2=0.81 RPD=2.20	The models map total sugars and dry matter content in each apple
Prediction of intact oranges consistency parameters	(Riccioli et al. 2021)	ANN	900–1,700 nmR_p^2	(SSC)$RMSE_{CV}$ = 0.87% (TA) $RMSE_{CV}$ =0.23 g L^- $RMSE_{CV}$ =2.78 for MI $RMSE_{CV}$ =1.11 for brima	ANN correlates hyperspectral NIR photos of orange chemical parameters effectively
Maturity determination of single maize seed	(Z. Wang, Tian et al. 2021)	Near-infrared HIS, partial least square discriminant analysis (PLS-DA Adaptive boosting (AdaBoost)	1,000–2,300 nm	Classification accuracy was 98.7%	Technology for the rapid and accurate classification of maize seed maturity

Variety identification of coated maize kernels	(C. Zhang et al. 2020)	LR, SVM, CNN, RNN, and LSTM	874–1,734 nm	Classification accuracy over 90%	Results demonstrated hyper-specific images from near-infrared fields with deep learning approaches to distinguish coated maize varieties
Corn leaf imager	(L. Wang et al. 2020)	Leafspec, GSI, HSI	–	Nitrogen content (N_2) and relative water content (H_2O) with $R^2_{N2} = 0.880$, $R^2 = 0.771$, $RMSE_{N2} = 0.265$ $RMSE_{H2O}$ 0.049	Low cost
Using a Cartesian robotic platform for HIS of corn plants	(Z. Chen et al. 2021)	Leafspec, planning algorithm	–	Average cycle time of 86 s. The R^2 value of 0.7307 Nitrogen treatments with P-values of 0.0193 and 0.0102	The machine engineer had similar results in field experiments with human operators
Predict sugar beet's compositions and mechanical properties	(Pan et al. 2016)	PLRE, UVE	500–1,000 nm	Correlations of 0.75–0.88 SEP = 0.95–1.08	Simpler algorithm
Variety discrimination of maize seeds	(S. Yang et al. 2017)	Unsupervised joint skewness-based wavelength selection algorithm, PCA, MNF, SMACC, NWHFC, FCLS,	924–1,657 nm	Classification accuracy = 96.57%	Simplex growing algorithm
Identification of wheat powdery mildew	(L. Huang et al. 2019)	SVM, develop a new vegetation index (NDVI1), LS-SVM, Relief-F algorithm	636 nm, 784 nm	Coefficient of determination (R^2) of 0.75–0.49	The study findings will provide a useful guideline for the calculation of wheat PM using the hyperspectral data on the field

REFERENCES

Abdel-Rahman, E. M., Mutanga, O., Odindi, J., Adam, E., Odindo, A., & Ismail, R. (2017). Estimating Swiss chard foliar macro- and micronutrient concentrations under different irrigation water sources using ground-based hyperspectral data and four partial least squares (PLS)-based (PLS1, PLS2, SPLS1 and SPLS2) regression algorithms. *Computers and Electronics in Agriculture, 132*, 21–33. https://doi.org/10.1016/j.compag.2016.11.008

Abdulridha, J., Ampatzidis, Y., Kakarla, S. C., & Roberts, P. (2020). Detection of target spot and bacterial spot diseases in tomato using UAV-based and benchtop-based hyperspectral imaging techniques. *Precision Agriculture, 21*(5), 955–978. https://doi.org/10.1007/s11119-019-09703-4

Al-Gaadi, K. A., Hassaballa, A. A., Tola, E., Kayad, A. G., Madugundu, R., Alblewi, B., & Assiri, F. (2016). Prediction of potato crop yield using precision agriculture techniques. *PLoS ONE, 11*(9), 1–16. https://doi.org/10.1371/journal.pone.0162219

Al Makdessi, N., Ecarnot, M., Roumet, P., & Rabatel, G. (2019). A spectral correction method for multi-scattering effects in close range hyperspectral imagery of vegetation scenes: Application to nitrogen content assessment in wheat. *Precision Agriculture, 20*(2), 237–259. https://doi.org/10.1007/s11119-018-9613-2

Arias, F., Zambrano, M., Broce, K., Medina, C., Pacheco, H., & Nunez, Y. (2020). Hyperspectral imaging for rice cultivation: Applications, methods and challenges. *AIMS Agriculture and Food, 6*(1), 273–307. https://doi.org/10.3934/AGRFOOD.2021018

Banerjee, S., & Shanmugam, P. (2021). Novel method for reconstruction of hyperspectral resolution images from multispectral data for complex coastal and inland waters. *Advances in Space Research, 67*(1), 266–289. https://doi.org/10.1016/j.asr.2020.09.045

Bangelesa, F., Adam, E., Knight, J., Dhau, I., Ramudzuli, M., & Mokotjomela, T. M. (2020). Predicting soil organic carbon content using hyperspectral remote sensing in a degraded mountain landscape in lesotho. *Applied and Environmental Soil Science, 2020*. https://doi.org/10.1155/2020/2158573

Barbin, D. F., Elmasry, G., Sun, D. W., & Allen, P. (2012a). Predicting quality and sensory attributes of pork using near-infrared hyperspectral imaging. *Analytica Chimica Acta, 719*, 30–42. https://doi.org/10.1016/j.aca.2012.01.004

Barbin, D. F., Elmasry, G., Sun, D. W., & Allen, P. (2012b). Near-infrared hyperspectral imaging for grading and classification of pork. *Meat Science, 90*(1), 259–268. https://doi.org/10.1016/j.meatsci.2011.07.011

Barbin, D. F., Elmasry, G., Sun, D. W., & Allen, P. (2013). Non-destructive determination of chemical composition in intact and minced pork using near-infrared hyperspectral imaging. *Food Chemistry, 138*(2–3), 1162–1171. https://doi.org/10.1016/j.foodchem.2012.11.120

Bianchini, V. de J. M., Mascarin, G. M., Silva, L. C. A. S., Arthur, V., Carstensen, J. M., Boelt, B., & Barboza da Silva, C. (2021). Multispectral and X-ray images for characterization of Jatropha curcas L. seed quality. *Plant Methods, 17*(1). https://doi.org/10.1186/s13007-021-00709-6

Bock, C. H., Poole, G. H., Parker, P. E., & Gottwald, T. R. (2010). Plant disease severity estimated visually, by digital photography and image analysis, and by hyperspectral imaging. *Critical Reviews in Plant Sciences, 29*(2), 59–107. https://doi.org/10.1080/07352681003617285

Caporaso, N., Whitworth, M. B., & Fisk, I. D. (2018, September 14). Near-Infrared spectroscopy and hyperspectral imaging for non-destructive quality assessment of cereal grains. *Applied Spectroscopy Reviews*. Taylor and Francis Inc. https://doi.org/10.1080/05704928.2018.1425214

Chang, C. (2003). *Hyperspectral Imaging: Techniques for Spectral Detection and Classification.* https://books.google.co.in/books?hl=en&lr=&id=JhBbXwFaA6sC&oi=fnd&pg=PA1&

dq=Hyperspectral+Imaging:+Techniques+for+Spectral+Detection+and+Classification.&ots=r3eEuUEXzW&sig=BBUG0SaTD1zdGCcFeofQe-MgXaA. Accessed 2 May 2021.

Chen, S., Zhang, F., Ning, J., Liu, X., Zhang, Z., & Yang, S. (2015). Predicting the anthocyanin content of wine grapes by NIR hyperspectral imaging. *Food Chemistry, 172*, 788–793. https://doi.org/10.1016/j.foodchem.2014.09.119

Chen, Y. R., Chao, K., & Kim, M. S. (2002). Machine vision technology for agricultural applications. *Computers and Electronics in Agriculture, 36*(2–3), 173–191. https://doi.org/10.1016/S0168-1699(02)00100-X

Chen, Y., Wang, J., Liu, G., Yang, Y., Liu, Z., & Deng, H. (2019). Hyperspectral estimation model of forest soil organic matter in Northwest Yunnan Province, China. *Forests, 10*(3), 217. https://doi.org/10.3390/f10030217

Chen, Z., Wang, J., Wang, T., Song, Z., Li, Y., Huang, Y., et al. (2021). Automated in-field leaf-level hyperspectral imaging of corn plants using a Cartesian robotic platform. *Computers and Electronics in Agriculture, 183*, 105996. https://doi.org/10.1016/j.compag.2021.105996

Costa, C., Antonucci, F., Pallottino, F., Aguzzi, J., Sun, D. W., & Menesatti, P. (2011). Shape analysis of agricultural products: A review of recent research advances and potential application to computer vision. *Food and Bioprocess Technology, 4*(5), 673–692. https://doi.org/10.1007/s11947-011-0556-0

Cubero, S., Aleixos, N., Moltó, E., Gómez-Sanchis, J., & Blasco, J. (2011). Advances in machine vision applications for automatic inspection and quality evaluation of fruits and vegetables. *Food and Bioprocess Technology, 4*(4), 487–504. https://doi.org/10.1007/s11947-010-0411-8

da Conceição, R. R. P., Simeone, M. L. F., Queiroz, V. A. V., de Medeiros, E. P., de Araújo, J. B., Coutinho, W. M., et al. (2021). Application of near-infrared hyperspectral (NIR) images combined with multivariate image analysis in the differentiation of two mycotoxicogenic Fusarium species associated with maize. *Food Chemistry, 344*, 128615. https://doi.org/10.1016/j.foodchem.2020.128615

Dixit, Y., Hitchman, S., Hicks, T. M., Lim, P., Wong, C. K., Holibar, L., et al. (2020). Non-invasive spectroscopic and imaging systems for prediction of beef quality in a meat processing pilot plant. *Meat Science*. https://doi.org/10.1016/j.meatsci.2020.108410

Du, L., Shi, S., Yang, J., Sun, J., & Gong, W. (2016). Using different regression methods to estimate leaf nitrogen content in rice by fusing hyperspectral LiDAR data and laser-induced chlorophyll fluorescence data. *Remote Sensing, 8*(6), 526. https://doi.org/10.3390/rs8060526

Du, Z., Zeng, X., Li, X., Ding, X., Cao, J., & Jiang, W. (2020). Recent advances in imaging techniques for bruise detection in fruits and vegetables. *Trends in Food Science and Technology, 99*, 133–141. https://doi.org/10.1016/j.tifs.2020.02.024

ElMasry, G., Sun, D. W., & Allen, P. (2013). Chemical-free assessment and mapping of major constituents in beef using hyperspectral imaging. *Journal of Food Engineering, 117*(2), 235–246. https://doi.org/10.1016/j.jfoodeng.2013.02.016

Eshkabilov, S., Lee, A., Sun, X., Lee, C. W., & Simsek, H. (2021). Hyperspectral imaging techniques for rapid detection of nutrient content of hydroponically grown lettuce cultivars. *Computers and Electronics in Agriculture, 181*, 105968. https://doi.org/10.1016/j.compag.2020.105968

Fact.MR. (2019). *Hyperspectral Imaging Market Forecast, Trend Analysis & Competition Tracking—Global Market Insights 2019 to 2029*. www.factmr.com/report/4570/hyperspectral-imaging-market. Accessed 1 May 2021.

Falkovskaya, A., & Gowen, A. (2020, July 1). Literature review: Spectral imaging applied to poultry products. *Poultry Science*. Elsevier Inc. https://doi.org/10.1016/j.psj.2020.04.013

Fei, B. (2020). Hyperspectral imaging in medical applications. In *Data Handling in Science and Technology* (Vol. 32, pp. 523–565). Elsevier Ltd. https://doi.org/10.1016/B978-0-444-63977-6.00021-3

Femenias, A., Bainotti, M. B., Gatius, F., Ramos, A. J., & Marín, S. (2021). Standardization of near infrared hyperspectral imaging for wheat single kernel sorting according to deoxynivalenol level. *Food Research International, 139.* https://doi.org/10.1016/j.foodres.2020.109925

Feng, L., Zhu, S., Liu, F., He, Y., Bao, Y., & Zhang, C. (2019, August 8). Hyperspectral imaging for seed quality and safety inspection: A review. *Plant Methods*. BioMed Central Ltd. https://doi.org/10.1186/s13007-019-0476-y

Forchetti, D. A. P., & Poppi, R. J. (2017). Use of NIR hyperspectral imaging and multivariate curve resolution (MCR) for detection and quantification of adulterants in milk powder. *LWT—Food Science and Technology*, *76*, 337–343. https://doi.org/10.1016/j.lwt.2016.06.046

Foster, A. J., Kakani, V. G., & Mosali, J. (2017). Estimation of bioenergy crop yield and N status by hyperspectral canopy reflectance and partial least square regression. *Precision Agriculture*, *18*(2), 192–209. https://doi.org/10.1007/s11119-016-9455-8

Fowler, S. M., Ponnampalam, E. N., Schmidt, H., Wynn, P., & Hopkins, D. L. (2015). Prediction of intramuscular fat content and major fatty acid groups of lamb M. longissimus lumborum using Raman spectroscopy. *Meat Science*, *110*, 70–75. https://doi.org/10.1016/j.meatsci.2015.06.016

Fox, G., & Manley, M. (2014). Applications of single kernel conventional and hyperspectral imaging near infrared spectroscopy in cereals. *Journal of the Science of Food and Agriculture*, *94*(2), 174–179. https://doi.org/10.1002/jsfa.6367

Gao, D., Li, M., Zhang, J., Song, D., Sun, H., Qiao, L., & Zhao, R. (2021). Improvement of chlorophyll content estimation on maize leaf by vein removal in hyperspectral image. *Computers and Electronics in Agriculture*, *184*, 106077. https://doi.org/10.1016/j.compag.2021.106077

Gao, Z., Khot, L. R., Naidu, R. A., & Zhang, Q. (2020). Early detection of grapevine leafroll disease in a red-berried wine grape cultivar using hyperspectral imaging. *Computers and Electronics in Agriculture*, *179*, 105807. https://doi.org/10.1016/j.compag.2020.105807

Geipel, J., Bakken, A. K., Jørgensen, M., & Korsaeth, A. (2021). Forage yield and quality estimation by means of UAV and hyperspectral imaging. *Precision Agriculture*, 1–27. https://doi.org/10.1007/s11119-021-09790-2

Gevaert, C. M., Suomalainen, J., Tang, J., & Kooistra, L. (2015). Generation of spectral-temporal response surfaces by combining multispectral satellite and hyperspectral UAV imagery for precision agriculture applications. *IEEE Journal of Selected Topics in Applied Earth Observations and Remote Sensing*, *8*(6), 3140–3146. https://doi.org/10.1109/JSTARS.2015.2406339

Goetz, A. F. H., Vane, G., Solomon, J. E., & Rock, B. N. (1985). Imaging spectrometry for earth remote sensing. *Science*, *228*(4704), 1147–1153. https://doi.org/10.1126/science.228.4704.1147

Gold, K. M., Townsend, P. A., Chlus, A., Herrmann, I., Couture, J. J., Larson, E. R., & Gevens, A. J. (2020). Hyperspectral measurements enable pre-symptomatic detection and differentiation of contrasting physiological effects of late blight and early blight in potato. *Remote Sensing*, *12*(2), 286. https://doi.org/10.3390/rs12020286

Gorretta, N., Nouri, M., Herrero, A., Gowen, A., & Roger, J. M. (2019). Early detection of the fungal disease "apple scab" using SWIR hyperspectral imaging. In *Workshop on Hyperspectral Image and Signal Processing, Evolution in Remote Sensing* (Vol. 2019-September). IEEE Computer Society. https://doi.org/10.1109/WHISPERS.2019.8921066

Gowen, A. A., O'Donnell, C. P., Taghizadeh, M., Cullen, P. J., Frias, J. M., & Downey, G. (2008). Hyperspectral imaging combined with principal component analysis for bruise damage detection on white mushrooms (Agaricus bisporus). *Journal of Chemometrics*, *22*(3–4), 259–267. https://doi.org/10.1002/cem.1127

Grafton, M., Kaul, T., Palmer, A., Bishop, P., & White, M. (2019). Technical note: Regression analysis of proximal hyperspectral data to predict soil pH and Olsen P. *Agriculture*, *9*(3), 55. https://doi.org/10.3390/agriculture9030055

Green, R. O., Eastwood, M. L., Sarture, C. M., Chrien, T. G., Aronsson, M., Chippendale, B. J., et al. (1998). Imaging spectroscopy and the Airborne Visible/Infrared Imaging Spectrometer (AVIRIS). *Remote Sensing of Environment*, *65*(3), 227–248. https://doi.org/10.1016/S0034-4257(98)00064-9

Guo, J., Zhang, J., Xiong, S., Zhang, Z., Wei, Q., Zhang, W., et al. (2021). Hyperspectral assessment of leaf nitrogen accumulation for winter wheat using different regression modeling. *Precision Agriculture*, 1–25. https://doi.org/10.1007/s11119-021-09804-z

Guo, L., Sun, X., Fu, P., Shi, T., Dang, L., Chen, Y., et al. (2021). Mapping soil organic carbon stock by hyperspectral and time-series multispectral remote sensing images in low-relief agricultural areas. *Geoderma*, *398*. https://doi.org/10.1016/j.geoderma.2021.115118

Guo, L., Zhang, H., Shi, T., Chen, Y., Jiang, Q., & Linderman, M. (2019). Prediction of soil organic carbon stock by laboratory spectral data and airborne hyperspectral images. *Geoderma*, *337*, 32–41. https://doi.org/10.1016/j.geoderma.2018.09.003

Gutiérrez, S., Wendel, A., & Underwood, J. (2019). Ground based hyperspectral imaging for extensive mango yield estimation. *Computers and Electronics in Agriculture*, *157*, 126–135. https://doi.org/10.1016/j.compag.2018.12.041

He, X., Yan, C., Jiang, X., Shen, F., You, J., & Fang, Y. (2021). Classification of aflatoxin B1 naturally contaminated peanut using visible and near-infrared hyperspectral imaging by integrating spectral and texture features. *Infrared Physics and Technology*, *114*, 103652. https://doi.org/10.1016/j.infrared.2021.103652

Hong, Y., Guo, L., Chen, S., Linderman, M., Mouazen, A. M., Yu, L., et al. (2020). Exploring the potential of airborne hyperspectral image for estimating topsoil organic carbon: Effects of fractional-order derivative and optimal band combination algorithm. *Geoderma*, *365*, 114228. https://doi.org/10.1016/j.geoderma.2020.114228

Hu, J., Peng, J., Zhou, Y., Xu, D., Zhao, R., Jiang, Q., et al. (2019). Quantitative estimation of soil salinity using UAV-borne hyperspectral and satellite multispectral images. *Remote Sensing*, *11*(7), 736. https://doi.org/10.3390/rs11070736

Hu, N., Li, W., Du, C., Zhang, Z., Gao, Y., Sun, Z., et al. (2021). Predicting micronutrients of wheat using hyperspectral imaging. *Food Chemistry*, *343*, 128473. https://doi.org/10.1016/j.foodchem.2020.128473

Huang, B., Yan, S., Xiao, L., Ji, R., Yang, L., Miao, A.-J., & Wang, P. (2018). Label-free imaging of nanoparticle uptake competition in single cells by hyperspectral stimulated Raman scattering. *Small*, *14*(10), 1703246. https://doi.org/10.1002/smll.201703246

Huang, L., Ding, W., Liu, W., Zhao, J., Huang, W., Xu, C., et al. (2019). Identification of wheat powdery mildew using in-situ hyperspectral data and linear regression and support vector machines. *Journal of Plant Pathology*, *101*(4), 1035–1045. https://doi.org/10.1007/s42161-019-00334-2

Ishida, T., Kurihara, J., Viray, F. A., Namuco, S. B., Paringit, E. C., Perez, G. J., et al. (2018). A novel approach for vegetation classification using UAV-based hyperspectral imaging. *Computers and Electronics in Agriculture*, *144*, 80–85. https://doi.org/10.1016/j.compag.2017.11.027

Jawaid, S., Talpur, F. N., Sherazi, S. T. H., Nizamani, S. M., & Khaskheli, A. A. (2013). Rapid detection of melamine adulteration in dairy milk by SB-ATR-Fourier transform infrared spectroscopy. *Food Chemistry*, *141*(3), 3066–3071. https://doi.org/10.1016/j.foodchem.2013.05.106

Ji, Y., Sun, L., Li, Y., Li, J., Liu, S., Xie, X., & Xu, Y. (2019). Non-destructive classification of defective potatoes based on hyperspectral imaging and support vector machine. *Infrared Physics and Technology*, *99*, 71–79. https://doi.org/10.1016/j.infrared.2019.04.007

Jiang, H., Ru, Y., Chen, Q., Wang, J., & Xu, L. (2021). Near-infrared hyperspectral imaging for detection and visualization of offal adulteration in ground pork. *Spectrochimica Acta—Part A: Molecular and Biomolecular Spectroscopy*, *249*. https://doi.org/10.1016/j.saa.2020.119307

Jin, X., Yang, G., Li, Z., Xu, X., Wang, J., & Lan, Y. (2018). Estimation of water productivity in winter wheat using the AquaCrop model with field hyperspectral data. *Precision Agriculture*, *19*(1), 1–17. https://doi.org/10.1007/s11119-016-9469-2

Kamruzzaman, M., Elmasry, G., Sun, D. W., & Allen, P. (2011). Application of NIR hyperspectral imaging for discrimination of lamb muscles. *Journal of Food Engineering*, *104*(3), 332–340. https://doi.org/10.1016/j.jfoodeng.2010.12.024

Khan, M. J., Khan, H. S., Yousaf, A., Khurshid, K., & Abbas, A. (2018, March 12). Modern trends in hyperspectral image analysis: A review. *IEEE Access*. Institute of Electrical and Electronics Engineers Inc. https://doi.org/10.1109/ACCESS.2018.2812999

Kim, D. M., Zhang, H., Zhou, H., Du, T., Wu, Q., Mockler, T. C., & Berezin, M. Y. (2015). Highly sensitive image-derived indices of water-stressed plants using hyperspectral imaging in SWIR and histogram analysis. *Scientific Reports*, *5*(1), 1–11. https://doi.org/10.1038/srep15919

Kimiya, T., Sivertsen, A. H., & Heia, K. (2013). VIS/NIR spectroscopy for non-destructive freshness assessment of Atlantic salmon (Salmo salar L.) fillets. *Journal of Food Engineering*, *116*(3), 758–764. https://doi.org/10.1016/j.jfoodeng.2013.01.008

Lan, W., Jaillais, B., Renard, C. M. G. C., Leca, A., Chen, S., Le Bourvellec, C., & Bureau, S. (2021). A method using near infrared hyperspectral imaging to highlight the internal quality of apple fruit slices. *Postharvest Biology and Technology*, *175*, 111497. https://doi.org/10.1016/j.postharvbio.2021.111497

Lassalle, G., Fabre, S., Credoz, A., Hédacq, R., Dubucq, D., & Elger, A. (2021). Mapping leaf metal content over industrial brownfields using airborne hyperspectral imaging and optimized vegetation indices. *Scientific Reports*, *11*(1), 1–13. https://doi.org/10.1038/s41598-020-79439-z

Li, B., Xu, X., Zhang, L., Han, J., Bian, C., Li, G., et al. (2020). Above-ground biomass estimation and yield prediction in potato by using UAV-based RGB and hyperspectral imaging. *ISPRS Journal of Photogrammetry and Remote Sensing*, *162*, 161–172. https://doi.org/10.1016/j.isprsjprs.2020.02.013

Li, W., Zhou, X., Yu, K., Zhang, Z., Liu, Y., Hu, N., et al. (2021). Spectroscopic estimation of N concentration in wheat organs for assessing N remobilization under different irrigation regimes. *Frontiers in Plant Science*, *12*. https://doi.org/10.3389/fpls.2021.657578

Li, X., Li, R., Wang, M., Liu, Y., Zhang, B., & Zhou, J. (2018). Hyperspectral imaging and their applications in the nondestructive quality assessment of fruits and vegetables. In *Hyperspectral Imaging in Agriculture, Food and Environment*. InTech. https://doi.org/10.5772/intechopen.72250

Li, Z., Nie, C., Wei, C., Xu, X., Song, X., & Wang, J. (2016). Comparison of four chemometric techniques for estimating leaf nitrogen concentrations in winter wheat (Triticum Aestivum) based on hyperspectral features. *Journal of Applied Spectroscopy*, *83*(2), 240–247. https://doi.org/10.1007/s10812-016-0276-3

Lim, J., Kim, G., Mo, C., Kim, M. S., Chao, K., Qin, J., et al. (2016). Detection of melamine in milk powders using near-infrared hyperspectral imaging combined with regression coefficient of partial least square regression model. *Talanta, 151*, 183–191. https://doi.org/10.1016/j.talanta.2016.01.035

Liu, C., Huang, W., Yang, G., Wang, Q., Li, J., & Chen, L. (2020). Determination of starch content in single kernel using near-infrared hyperspectral images from two sides of corn seeds. *Infrared Physics and Technology, 110*, 103462. https://doi.org/10.1016/j.infrared.2020.103462

Liu, S., Liu, X., Liu, M., Wu, L., Ding, C., & Huang, Z. (2017). Extraction of rice phenological differences under heavy metal stress using EVI time-series from HJ-1A/B data. *Sensors, 17*(6), 1243. https://doi.org/10.3390/s17061243

Liu, Z. Y., Qi, J. G., Wang, N. N., Zhu, Z. R., Luo, J., Liu, L. J., et al. (2018). Hyperspectral discrimination of foliar biotic damages in rice using principal component analysis and probabilistic neural network. *Precision Agriculture, 19*(6), 973–991. https://doi.org/10.1007/s11119-018-9567-4

Lowe, A., Harrison, N., & French, A. P. (2017). Hyperspectral image analysis techniques for the detection and classification of the early onset of plant disease and stress. *Plant Methods, 13*(1), 1–12. https://doi.org/10.1186/s13007-017-0233-z

Lu, Jingshan, Li, W., Yu, M., Zhang, X., Ma, Y., Su, X., et al. (2021). Estimation of rice plant potassium accumulation based on non-negative matrix factorization using hyperspectral reflectance. *Precision Agriculture, 22*(1), 51–74. https://doi.org/10.1007/s11119-020-09729-z

Lu, Jingshan, Yang, T., Su, X., Qi, H., Yao, X., Cheng, T., et al. (2020). Monitoring leaf potassium content using hyperspectral vegetation indices in rice leaves. *Precision Agriculture, 21*(2), 324–348. https://doi.org/10.1007/s11119-019-09670-w

Lu, Jinzhu, Zhou, M., Gao, Y., & Jiang, H. (2018). Using hyperspectral imaging to discriminate yellow leaf curl disease in tomato leaves. *Precision Agriculture, 19*(3), 379–394. https://doi.org/10.1007/s11119-017-9524-7

Ma, C., Ren, Z., Zhang, Z., Du, J., Jin, C., & Yin, X. (2021). Development of simplified models for nondestructive testing of rice (with husk) protein content using hyperspectral imaging technology. *Vibrational Spectroscopy, 114*, 103230. https://doi.org/10.1016/j.vibspec.2021.103230

Ma, D., Maki, H., Neeno, S., Zhang, L., Wang, L., & Jin, J. (2020). Application of non-linear partial least squares analysis on prediction of biomass of maize plants using hyperspectral images. *Biosystems Engineering, 200*, 40–54. https://doi.org/10.1016/j.biosystemseng.2020.09.002

Ma, J., Sun, D.-W., Pu, H., Cheng, J.-H., & Wei, Q. (2019). Advanced techniques for hyperspectral imaging in the food industry: Principles and recent applications. *Annual Review of Food Science and Technology, 10*(1), 197–220. https://doi.org/10.1146/annurev-food-032818-121155

Mahajan, G. R., Pandey, R. N., Sahoo, R. N., Gupta, V. K., Datta, S. C., & Kumar, D. (2017). Monitoring nitrogen, phosphorus and sulphur in hybrid rice (Oryza sativa L.) using hyperspectral remote sensing. *Precision Agriculture, 18*(5), 736–761. https://doi.org/10.1007/s11119-016-9485-2

Mahlein, A.-K., Alisaac, E., Al Masri, A., Behmann, J., Dehne, H.-W., & Oerke, E.-C. (2019). Comparison and combination of thermal, fluorescence, and hyperspectral imaging for monitoring fusarium head blight of wheat on spikelet scale. *Sensors, 19*(10), 2281. https://doi.org/10.3390/s19102281

Manley, M., Williams, P., Nilsson, D., & Geladi, P. (2009). Near infrared hyperspectral imaging for the evaluation of endosperm texture in whole yellow maize (Zea maize L.) kernels. *Journal of Agricultural and Food Chemistry, 57*(19), 8761–8769. https://doi.org/10.1021/jf9018323

Manolakis, D., Lockwood, R., & Cooley, T. (2016). *Hyperspectral Imaging Remote Sensing: Physics, Sensors, and Algorithms.* https://books.google.co.in/books?hl=en&lr=&id=rssi

DQAAQBAJ&oi=fnd&pg=PR11&ots=ozzGTnaV2l&sig=HE1-KIVWY82-vUtfMqnFvirE_5o. Accessed 2 May 2021.

Menesatti, P., Costa, C., & Aguzzi, J. (2010). Quality evaluation of fish by hyperspectral imaging. In *Hyperspectral Imaging for Food Quality Analysis and Control* (pp. 273–294). Elsevier Inc. https://doi.org/10.1016/B978-0-12-374753-2.10008-5

Moghadam, P., Ward, D., Goan, E., Jayawardena, S., Sikka, P., & Hernandez, E. (2017). Plant disease detection using hyperspectral imaging. In *DICTA 2017–2017 International Conference on Digital Image Computing: Techniques and Applications* (Vol. 2017-December, pp. 1–8). Institute of Electrical and Electronics Engineers Inc. https://doi.org/10.1109/DICTA.2017.8227476

Mozgeris, G., Juodkienė, V., Jonikavičius, D., Straigytė, L., Gadal, S., & Ouerghemmi, W. (2018). Ultra-light aircraft-based hyperspectral and colour-infrared imaging to identify deciduous tree species in an urban environment. *Remote Sensing*, *10*(10), 1668. https://doi.org/10.3390/rs10101668

Murphy, R. J., Whelan, B., Chlingaryan, A., & Sukkarieh, S. (2019). Quantifying leaf-scale variations in water absorption in lettuce from hyperspectral imagery: A laboratory study with implications for measuring leaf water content in the context of precision agriculture. *Precision Agriculture*, *20*(4), 767–787. https://doi.org/10.1007/s11119-018-9610-5

Naganathan, G. K., Grimes, L. M., Subbiah, J., Calkins, C. R., Samal, A., & Meyer, G. E. (2008). Visible/near-infrared hyperspectral imaging for beef tenderness prediction. *Computers and Electronics in Agriculture*, *64*(2), 225–233. https://doi.org/10.1016/j.compag.2008.05.020

Nagasubramanian, K., Jones, S., Sarkar, S., Singh, A. K., Singh, A., & Ganapathysubramanian, B. (2018). Hyperspectral band selection using genetic algorithm and support vector machines for early identification of charcoal rot disease in soybean stems. *Plant Methods*, *14*(1), 86. https://doi.org/10.1186/s13007-018-0349-9

Nagasubramanian, K., Jones, S., Singh, A. K., Sarkar, S., Singh, A., & Ganapathysubramanian, B. (2019). Plant disease identification using explainable 3D deep learning on hyperspectral images. *Plant Methods*, *15*(1), 1–10. https://doi.org/10.1186/s13007-019-0479-8

Naik, B. B., Naveen, H. R., Sreenivas, G., Choudary, K. K., Devkumar, D., & Adinarayana, J. (2020). Identification of water and nitrogen stress indicative spectral bands using hyperspectral remote sensing in maize during post-monsoon season. *Journal of the Indian Society of Remote Sensing*, *48*(12), 1787–1795. https://doi.org/10.1007/s12524-020-01200-w

Nakariyakul, S., & Casasent, D. P. (2009). Fast feature selection algorithm for poultry skin tumor detection in hyperspectral data. *Journal of Food Engineering*, *94*(3–4), 358–365. https://doi.org/10.1016/j.jfoodeng.2009.04.001

Nandibewoor, A., & Hegadi, R. (2019). A novel SMLR-PSO model to estimate the chlorophyll content in the crops using hyperspectral satellite images. *Cluster Computing*, *22*(1), 443–450. https://doi.org/10.1007/s10586-018-2243-7

Nguyen Do Trong, N., Tsuta, M., Nicolaï, B. M., De Baerdemaeker, J., & Saeys, W. (2011). Prediction of optimal cooking time for boiled potatoes by hyperspectral imaging. *Journal of Food Engineering*, *105*(4), 617–624. https://doi.org/10.1016/j.jfoodeng.2011.03.031

Nguyen, H. D. D., Pan, V., Pham, C., Valdez, R., Doan, K., & Nansen, C. (2020). Night-based hyperspectral imaging to study association of horticultural crop leaf reflectance and nutrient status. *Computers and Electronics in Agriculture*, *173*, 105458. https://doi.org/10.1016/j.compag.2020.105458

Nogales-Bueno, J., Baca-Bocanegra, B., Rodríguez-Pulido, F. J., Heredia, F. J., & Hernández-Hierro, J. M. (2015). Use of near infrared hyperspectral tools for the screening of extractable polyphenols in red grape skins. *Food Chemistry, 172*, 559–564. https://doi.org/10.1016/j.foodchem.2014.09.112

Nogales-Bueno, J., Rodríguez-Pulido, F. J., Heredia, F. J., & Hernández-Hierro, J. M. (2015). Comparative study on the use of anthocyanin profile, color image analysis and near-infrared hyperspectral imaging as tools to discriminate between four autochthonous red grape cultivars from la Rioja (Spain). *Talanta, 131*, 412–416. https://doi.org/10.1016/j.talanta.2014.07.086

Nyalala, I., Okinda, C., Kunjie, C., Korohou, T., Nyalala, L., & Chao, Q. (2021, May 1). Weight and volume estimation of poultry and products based on computer vision systems: A review. *Poultry Science*. Elsevier Inc. https://doi.org/10.1016/j.psj.2021.101072

Onoyama, H., Ryu, C., Suguri, M., & Iida, M. (2018). Estimation of rice protein content before harvest using ground-based hyperspectral imaging and region of interest analysis. *Precision Agriculture, 19*(4), 721–734. https://doi.org/10.1007/s11119-017-9552-3

Paliwal, J., Thakur, S., & Erkinbaev, C. (2018). Protein-starch interactions in cereal grains and pulses. In *Encyclopedia of Food Chemistry* (pp. 446–452). Elsevier. https://doi.org/10.1016/B978-0-08-100596-5.22349-4

Pallottino, F., Stazi, S. R., D'Annibale, A., Marabottini, R., Allevato, E., Antonucci, F., et al. (2018). Rapid assessment of As and other elements in naturally-contaminated calcareous soil through hyperspectral VIS-NIR analysis. *Talanta, 190*, 167–173. https://doi.org/10.1016/j.talanta.2018.07.082

Pan, L., Lu, R., Zhu, Q., Tu, K., & Cen, H. (2016). Predict compositions and mechanical properties of sugar beet using hyperspectral scattering. *Food and Bioprocess Technology, 9*(7), 1177–1186. https://doi.org/10.1007/s11947-016-1710-5

Pandey, P., Ge, Y., Stoerger, V., & Schnable, J. C. (2017). High throughput in vivo analysis of plant leaf chemical properties using hyperspectral imaging. *Frontiers in Plant Science, 8*, 1348. https://doi.org/10.3389/fpls.2017.01348

Pang, L., Men, S., Yan, L., & Xiao, J. (2020). Rapid vitality estimation and prediction of corn seeds based on spectra and images using deep learning and hyperspectral imaging techniques. *IEEE Access, 8*, 123026–123036. https://doi.org/10.1109/ACCESS.2020.3006495

Parfitt, J., Barthel, M., & MacNaughton, S. (2010). Food waste within food supply chains: Quantification and potential for change to 2050. *Philosophical Transactions of the Royal Society B: Biological Sciences, 365*(1554), 3065–3081. https://doi.org/10.1098/rstb.2010.0126

Park, B., Windham, W. R., Lawrence, K. C., & Smith, D. P. (2007). Contaminant classification of poultry hyperspectral imagery using a spectral angle mapper algorithm. *Biosystems Engineering, 96*(3), 323–333. https://doi.org/10.1016/j.biosystemseng.2006.11.012

Patel, A. K., & Ghosh, J. K. (2019). Soil fertility status assessment using hyperspectral remote sensing. *Remote Sensing for Agriculture, Ecosystems, and Hydrology XXI, 11149*, 14. https://doi.org/10.1117/12.2533115

Patel, A. K., Ghosh, J. K., Pande, S., & Sayyad, S. U. (2020). Deep-learning-based approach for estimation of fractional abundance of nitrogen in soil from hyperspectral data. *IEEE Journal of Selected Topics in Applied Earth Observations and Remote Sensing, 13*, 6495–6511. https://doi.org/10.1109/JSTARS.2020.3039844

Pearlman, J. S., Barry, P. S., Segal, C. C., Shepanski, J., Beiso, D., & Carman, S. L. (2003). Hyperion, a space-based imaging spectrometer. *IEEE Transactions on Geoscience and Remote Sensing, 41*(6 PART I), 1160–1173. https://doi.org/10.1109/TGRS.2003.815018

Peng, Y., Zhu, X., Xiong, J., Yu, R., Liu, T., Jiang, Y., & Yang, G. (2021). Estimation of nitrogen content on apple tree canopy through red-edge parameters from fractional-order differential operators using hyperspectral reflectance. *Journal of the Indian Society of Remote Sensing, 49*(2), 377–392. https://doi.org/10.1007/s12524-020-01197-2

Persistence Market Research. (2016). *Imaging Technology For Precision Agriculture Market.* www.persistencemarketresearch.com/market-research/imaging-technology-for-precision-agriculture-market.asp. Accessed 28 April 2021.

Pu, H., Sun, D. W., Ma, J., Liu, D., & Kamruzzaman, M. (2014). Hierarchical variable selection for predicting chemical constituents in lamb meats using hyperspectral imaging. *Journal of Food Engineering, 143*, 44–52. https://doi.org/10.1016/j.jfoodeng.2014.06.025

Qiu, G., Lü, E., Lu, H., Xu, S., Zeng, F., & Shui, Q. (2018). Single-kernel FT-NIR spectroscopy for detecting supersweet corn (Zea mays L. saccharata sturt) seed viability with multivariate data analysis. *Sensors (Switzerland), 18*(4). https://doi.org/10.3390/s18041010

Rabanera, J. D., Guzman, J. D., & Yaptenco, K. F. (2021). Rapid and Non-destructive measurement of moisture content of peanut (Arachis hypogaea L.) kernel using a near-infrared hyperspectral imaging technique. *Journal of Food Measurement and Characterization.* https://doi.org/10.1007/s11694-021-00894-x

Rady, A., Ekramirad, N., Adedeji, A. A., Li, M., & Alimardani, R. (2017). Hyperspectral imaging for detection of codling moth infestation in GoldRush apples. *Postharvest Biology and Technology, 129*, 37–44. https://doi.org/10.1016/j.postharvbio.2017.03.007

Rahman, A., Lee, H., Kim, M. S., & Cho, B. K. (2018). Mapping the pungency of green pepper using hyperspectral imaging. *Food Analytical Methods, 11*(11), 3042–3052. https://doi.org/10.1007/s12161-018-1275-1

Rathna Priya, T. S., & Manickavasagan, A. (2021). Characterising corn grain using infrared imaging and spectroscopic techniques: A review. *Journal of Food Measurement and Characterization.* Springer. https://doi.org/10.1007/s11694-021-00898-7

Ravikanth, L., Chelladurai, V., Jayas, D. S., & White, N. D. G. (2016). Detection of broken kernels content in bulk wheat samples using near-infrared hyperspectral imaging. *Agricultural Research, 5*(3), 285–292. https://doi.org/10.1007/s40003-016-0227-5

Reis, A. S., Rodrigues, M., Alemparte Abrantes dos Santos, G. L., Mayara de Oliveira, K., Furlanetto, R. H., Teixeira Crusiol, L. G., et al. (2021). Detection of soil organic matter using hyperspectral imaging sensor combined with multivariate regression modeling procedures. *Remote Sensing Applications: Society and Environment, 22*, 100492. https://doi.org/10.1016/j.rsase.2021.100492

Riccioli, C., Pérez-Marín, D., & Garrido-Varo, A. (2021). Optimizing spatial data reduction in hyperspectral imaging for the prediction of quality parameters in intact oranges. *Postharvest Biology and Technology, 176*, 111504. https://doi.org/10.1016/j.postharvbio.2021.111504

Rubio-Delgado, J., Pérez, C. J., & Vega-Rodríguez, M. A. (2021). Predicting leaf nitrogen content in olive trees using hyperspectral data for precision agriculture. *Precision Agriculture, 22*(1), 1–21. https://doi.org/10.1007/s11119-020-09727-1

Schmid, T., Rodriguez-Rastrero, M., Escribano, P., Palacios-Orueta, A., Ben-Dor, E., Plaza, A., et al. (2016). Characterization of soil erosion indicators using hyperspectral data from a Mediterranean rainfed cultivated region. *IEEE Journal of Selected Topics in Applied Earth Observations and Remote Sensing, 9*(2), 845–860. https://doi.org/10.1109/JSTARS.2015.2462125

Sendin, K., Manley, M., Baeten, V., Fernández Pierna, J. A., & Williams, P. J. (2019). Near infrared hyperspectral imaging for white maize classification according to grading regulations. *Food Analytical Methods, 12*(7), 1612–1624. https://doi.org/10.1007/s12161-019-01464-0

Sendin, K., Williams, P. J., & Manley, M. (2018). Near infrared hyperspectral imaging in quality and safety evaluation of cereals. *Critical Reviews in Food Science and Nutrition, 58*(4), 575–590. https://doi.org/10.1080/10408398.2016.1205548

Serranti, S., Trella, A., Bonifazi, G., & Izquierdo, C. G. (2018). Production of an innovative biowaste-derived fertilizer: Rapid monitoring of physical-chemical parameters by hyperspectral imaging. *Waste Management*, *75*, 141–148. https://doi.org/10.1016/j.wasman.2018.02.013

Shen, Q., Xia, K., Zhang, S., Kong, C., Hu, Q., & Yang, S. (2019). Hyperspectral indirect inversion of heavy-metal copper in reclaimed soil of iron ore area. *Spectrochimica Acta—Part A: Molecular and Biomolecular Spectroscopy*, *222*, 117191. https://doi.org/10.1016/j.saa.2019.117191

Shi, T., Liu, H., Chen, Y., Wang, J., & Wu, G. (2016). Estimation of arsenic in agricultural soils using hyperspectral vegetation indices of rice. *Journal of Hazardous Materials*, *308*, 243–252. https://doi.org/10.1016/j.jhazmat.2016.01.022

Silva, L. C. R., Folli, G. S., Santos, L. P., Barros, I. H. A. S., Oliveira, B. G., Borghi, F. T., et al. (2020). Quantification of beef, pork, and chicken in ground meat using a portable NIR spectrometer. *Vibrational Spectroscopy*, *111*. https://doi.org/10.1016/j.vibspec.2020.103158

Song, X., Xu, D., He, L., Feng, W., Wang, Y., Wang, Z., et al. (2016). Using multi-angle hyperspectral data to monitor canopy leaf nitrogen content of wheat. *Precision Agriculture*, *17*(6), 721–736. https://doi.org/10.1007/s11119-016-9445-x

Song, Y.-Q., Zhao, X., Su, H.-Y., Li, B., Hu, Y.-M., & Cui, X.-S. (2018). Predicting spatial variations in soil nutrients with hyperspectral remote sensing at regional scale. *Sensors*, *18*(9), 3086. https://doi.org/10.3390/s18093086

Stuart, M. B., McGonigle, A. J. S., & Willmott, J. R. (2019). Hyperspectral imaging in environmental monitoring: A review of recent developments and technological advances in compact field deployable systems. *Sensors*, *19*(14), 3071. https://doi.org/10.3390/s19143071

Suarez, L. A., Robson, A., McPhee, J., O'Halloran, J., & van Sprang, C. (2020). Accuracy of carrot yield forecasting using proximal hyperspectral and satellite multispectral data. *Precision Agriculture*, *21*(6), 1304–1326. https://doi.org/10.1007/s11119-020-09722-6

Sun, D., Cen, H., Weng, H., Wan, L., Abdalla, A., El-Manawy, A. I., et al. (2019). Using hyperspectral analysis as a potential high throughput phenotyping tool in GWAS for protein content of rice quality. *Plant Methods*, *15*(1), 1–16. https://doi.org/10.1186/s13007-019-0432-x

Sun, Y., Liu, Y., Yu, H., Xie, A., Li, X., Yin, Y., & Duan, X. (2017). Non-destructive prediction of moisture content and freezable water content of purple-fleshed sweet potato slices during drying process using hyperspectral imaging technique. *Food Analytical Methods*, *10*(5), 1535–1546. https://doi.org/10.1007/s12161-016-0722-0

Susič, N., Žibrat, U., Širca, S., Strajnar, P., Razinger, J., Knapič, M., et al. (2018). Discrimination between abiotic and biotic drought stress in tomatoes using hyperspectral imaging. *Sensors and Actuators, B: Chemical*, *273*, 842–852. https://doi.org/10.1016/j.snb.2018.06.121

Tahmasbian, I., Morgan, N. K., Bai, S. H., Dunlop, M. W., & Moss, A. F. (2021). Comparison of hyperspectral imaging and near-infrared spectroscopy to determine nitrogen and carbon concentrations in wheat. *Remote Sensing*, *13*(6). https://doi.org/10.3390/rs13061128

Tan, K., Wang, H., Chen, L., Du, Q., Du, P., & Pan, C. (2020). Estimation of the spatial distribution of heavy metal in agricultural soils using airborne hyperspectral imaging and random forest. *Journal of Hazardous Materials*, *382*, 120987. https://doi.org/10.1016/j.jhazmat.2019.120987

Tan, K., Wang, X., Zhu, J., Hu, J., & Li, J. (2018). A novel active learning approach for the classification of hyperspectral imagery using quasi-Newton multinomial logistic regression. *International Journal of Remote Sensing*, *39*(10), 3029–3054. https://doi.org/10.1080/01431161.2018.1433893

Tian, S., Wang, S., Bai, X., Zhou, D., Lu, Q., Wang, M., & Wang, J. (2020). Hyperspectral estimation model of soil Pb content and its applicability in different soil types. *Acta Geochimica*, *39*(3), 423–433. https://doi.org/10.1007/s11631-019-00388-0

Tian, Y. C., Yao, X., Yang, J., Cao, W. X., Hannaway, D. B., & Zhu, Y. (2011). Assessing newly developed and published vegetation indices for estimating rice leaf nitrogen concentration with ground- and space-based hyperspectral reflectance. *Field Crops Research*, *120*(2), 299–310. https://doi.org/10.1016/j.fcr.2010.11.002

Tong, X., Duan, L., Liu, T., & Singh, V. P. (2019). Combined use of in situ hyperspectral vegetation indices for estimating pasture biomass at peak productive period for harvest decision. *Precision Agriculture*, *20*(3), 477–495. https://doi.org/10.1007/s11119-018-9592-3

Vanegas, F., Bratanov, D., Powell, K., Weiss, J., & Gonzalez, F. (2018). A novel methodology for improving plant pest surveillance in vineyards and crops using UAV-based hyperspectral and spatial data. *Sensors*, *18*(1), 260. https://doi.org/10.3390/s18010260

Wakholi, C., Kandpal, L. M., Lee, H., Bae, H., Park, E., Kim, M. S., et al. (2018). Rapid assessment of corn seed viability using short wave infrared line-scan hyperspectral imaging and chemometrics. *Sensors and Actuators, B: Chemical*, *255*, 498–507. https://doi.org/10.1016/j.snb.2017.08.036

Wang, G., Wang, Q., Su, Z., & Zhang, J. (2020). Predicting copper contamination in wheat canopy during the full growth period using hyperspectral data. *Environmental Science and Pollution Research*, *27*(31), 39029–39040. https://doi.org/10.1007/s11356-020-09973-w

Wang, J., Zhang, C., Shi, Y., Long, M., Islam, F., Yang, C., et al. (2020). Evaluation of quinclorac toxicity and alleviation by salicylic acid in rice seedlings using ground-based visible/near-infrared hyperspectral imaging. *Plant Methods*, *16*(1). https://doi.org/10.1186/s13007-020-00576-7

Wang, L., Jin, J., Song, Z., Wang, J., Zhang, L., Rehman, T. U., et al. (2020). LeafSpec: An accurate and portable hyperspectral corn leaf imager. *Computers and Electronics in Agriculture*, *169*, 105209. https://doi.org/10.1016/j.compag.2019.105209

Wang, Y., Hu, X., Hou, Z., Ning, J., & Zhang, Z. (2018). Discrimination of nitrogen fertilizer levels of tea plant (*Camellia sinensis*) based on hyperspectral imaging. *Journal of the Science of Food and Agriculture*, *98*(12), 4659–4664. https://doi.org/10.1002/jsfa.8996

Wang, Z., Fan, S., Wu, J., Zhang, C., Xu, F., Yang, X., & Li, J. (2021). Application of long-wave near infrared hyperspectral imaging for determination of moisture content of single maize seed. *Spectrochimica Acta—Part A: Molecular and Biomolecular Spectroscopy*, *254*, 119666. https://doi.org/10.1016/j.saa.2021.119666

Wang, Z., Tian, X., Fan, S., Zhang, C., & Li, J. (2021). Maturity determination of single maize seed by using near-infrared hyperspectral imaging coupled with comparative analysis of multiple classification models. *Infrared Physics and Technology*, *112*, 103596. https://doi.org/10.1016/j.infrared.2020.103596

Wang, Z., Zhang, Y., Fan, S., Jiang, Y., & Li, J. (2020). Determination of moisture content of single maize seed by using long-wave near-infrared hyperspectral imaging (LWNIR) Coupled with UVE-SPA combination variable selection method. *IEEE Access*, *8*, 195229–195239. https://doi.org/10.1109/ACCESS.2020.3033582

Wei, L., Wang, Z., Huang, C., Zhang, Y., Wang, Z., Xia, H., & Cao, L. (2020). Transparency estimation of narrow rivers by UAV-borne hyperspectral remote sensing imagery. *IEEE Access*, *8*, 168137–168153. https://doi.org/10.1109/ACCESS.2020.3023690

Wen, P., Shi, Z., Li, A., Ning, F., Zhang, Y., Wang, R., & Li, J. (2020). Estimation of the vertically integrated leaf nitrogen content in maize using canopy hyperspectral red edge parameters. *Precision Agriculture*, *22*(3), 984–1005. https://doi.org/10.1007/s11119-020-09769-5

Wu, D., & Sun, D. W. (2013). Application of visible and near infrared hyperspectral imaging for non-invasively measuring distribution of water-holding capacity in salmon flesh. *Talanta*, *116*, 266–276. https://doi.org/10.1016/j.talanta.2013.05.030

Xie, A., Sun, D. W., Xu, Z., & Zhu, Z. (2015). Rapid detection of frozen pork quality without thawing by Vis-NIR hyperspectral imaging technique. *Talanta*, *139*, 208–215. https://doi.org/10.1016/j.talanta.2015.02.027

Xing, F., Yao, H., Liu, Y., Dai, X., Brown, R. L., & Bhatnagar, D. (2019, January 2). Recent developments and applications of hyperspectral imaging for rapid detection of mycotoxins and mycotoxigenic fungi in food products. *Critical Reviews in Food Science and Nutrition*. Taylor and Francis Inc. https://doi.org/10.1080/10408398.2017.1363709

Yan, Y., & Yu, W. (2019). Early detection of rice blast (Pyricularia) at seedling stage based on near-infrared hyper-spectral image. In *ACM International Conference Proceeding Series* (pp. 64–68). Association for Computing Machinery. https://doi.org/10.1145/3369166.3369185

Yang, S., Zhu, Q. B., Huang, M., & Qin, J. W. (2017). Hyperspectral image-based variety discrimination of maize seeds by using a multi-model strategy coupled with unsupervised joint skewness-based wavelength selection algorithm. *Food Analytical Methods*, *10*(2), 424–433. https://doi.org/10.1007/s12161-016-0597-0

Yang, W., Nigon, T., Hao, Z., Dias Paiao, G., Fernández, F. G., Mulla, D., & Yang, C. (2021). Estimation of corn yield based on hyperspectral imagery and convolutional neural network. *Computers and Electronics in Agriculture*, *184*, 106092. https://doi.org/10.1016/j.compag.2021.106092

Yang, W., Yang, C., Hao, Z., Xie, C., & Li, M. (2019). Diagnosis of plant cold damage based on hyperspectral imaging and convolutional neural network. *IEEE Access*, *7*, 118239–118248. https://doi.org/10.1109/ACCESS.2019.2936892

Ye, X., Abe, S., & Zhang, S. (2020). Estimation and mapping of nitrogen content in apple trees at leaf and canopy levels using hyperspectral imaging. *Precision Agriculture*, *21*(1), 198–225. https://doi.org/10.1007/s11119-019-09661-x

Zaeem, M., Nadeem, M., Huong Pham, T., Ashiq, W., Ali, W., Shah Mohioudin Gillani, S., et al. (2021). Development of a hyperspectral imaging technique using LA-ICP-MS to show the spatial distribution of elements in soil cores. *Geoderma*, *385*, 114831. https://doi.org/10.1016/j.geoderma.2020.114831

Zeng, F., Lü, E., Qiu, G., Lu, H., & Jiang, B. (2019). Single-kernel ft-NIR spectroscopy for detecting maturity of cucumber seeds using a multiclass hierarchical classification strategy. *Applied Sciences (Switzerland)*, *9*(23). https://doi.org/10.3390/app9235058

Zhang, B., Huang, W., Li, J., Zhao, C., Fan, S., Wu, J., & Liu, C. (2014, August 1). Principles, developments and applications of computer vision for external quality inspection of fruits and vegetables: A review. *Food Research International*. Elsevier Ltd. https://doi.org/10.1016/j.foodres.2014.03.012

Zhang, C., Jiang, H., Liu, F., & He, Y. (2017). Application of near-infrared hyperspectral imaging with variable selection methods to determine and visualize caffeine content of coffee beans. *Food and Bioprocess Technology*, *10*(1), 213–221. https://doi.org/10.1007/s11947-016-1809-8

Zhang, C., Zhao, Y., Yan, T., Bai, X., Xiao, Q., Gao, P., et al. (2020). Application of near-infrared hyperspectral imaging for variety identification of coated maize kernels with deep learning. *Infrared Physics and Technology*, *111*, 103550. https://doi.org/10.1016/j.infrared.2020.103550

Zhang, G. S., Xu, T. Y., Tian, Y. W., Xu, H., Song, J. Y., & Lan, Y. (2020). Assessment of rice leaf blast severity using hyperspectral imaging during late vegetative growth. *Australasian Plant Pathology*, *49*(5), 571–578. https://doi.org/10.1007/s13313-020-00736-2

Zhang, H., Zhang, S., Chen, Y., Luo, W., Huang, Y., Tao, D., et al. (2020). Non-destructive determination of fat and moisture contents in Salmon (Salmo salar) fillets using near-infrared hyperspectral imaging coupled with spectral and textural features. *Journal of Food Composition and Analysis*, *92*. https://doi.org/10.1016/j.jfca.2020.103567

Zhang, J., Dai, L., & Cheng, F. (2021a). Identification of corn seeds with different freezing damage degree based on hyperspectral reflectance imaging and deep learning method. *Food Analytical Methods*, *14*(2), 389–400. https://doi.org/10.1007/s12161-020-01871-8

Zhang, J., Dai, L., & Cheng, F. (2021b). Corn seed variety classification based on hyperspectral reflectance imaging and deep convolutional neural network. *Journal of Food Measurement and Characterization*, *15*(1), 484–494. https://doi.org/10.1007/s11694-020-00646-3

Zhang, S., Shen, Q., Nie, C., Huang, Y., Wang, J., Hu, Q., et al. (2019). Hyperspectral inversion of heavy metal content in reclaimed soil from a mining wasteland based on different spectral transformation and modeling methods. *Spectrochimica Acta—Part A: Molecular and Biomolecular Spectroscopy*, *211*, 393–400. https://doi.org/10.1016/j.saa.2018.12.032

Zheng, H., Cheng, T., Li, D., Zhou, X., Yao, X., Tian, Y., et al. (2018). Evaluation of RGB, color-infrared and multispectral images acquired from unmanned aerial systems for the estimation of nitrogen accumulation in rice. *Remote Sensing*, *10*(6), 824. https://doi.org/10.3390/rs10060824

Zovko, M., Žibrat, U., Knapič, M., Kovačić, M. B., & Romić, D. (2019). Hyperspectral remote sensing of grapevine drought stress. *Precision Agriculture*, *20*(2), 335–347. https://doi.org/10.1007/s11119-019-09640-2

2 Early Prediction of COVID-19 Using Modified Convolutional Neural Networks

Asadi Srinivasulu, Umesh Neelakantan, Tarkeswar Barua, Srinivas Nowduri, and MM Subramanyam

CONTENTS

DOI: 10.1201/9781003152392-2

2.1 INTRODUCTION

Corona is a group of viruses that cause disease in mammals. COVID-19 (Co from corona, Vi from Virus, and D stands for Disease, and it was introduced by World Health Organization [WHO] on December 31, 2019) is one of the corona family. The very prime victim was confirmed in China—Wuhan city (Guangdong state), also known as Wu Flue. COVID-19 has got the longest genome of RNA (Robo Nucleic Acid), approximately 26–32k long. COVID-19 malignant might be the most widely recognized kind of disease among inquest folks in the entire world (Botta A. et al. 2016). In 2017, it had been the third significant reason for biting the dust from COVID-19 in men in the United States, with around 161,360 most recent cases which represented 19% of most new disease occurrences and 26,730 fatalities, which showed 8% of most malignant growth passing (Cheng J.-Z. et al. 2016). Despite the fact that COVID-19 might be defined as basically being dangerous, whenever found in the main stages, the achievement rates are exorbitant due to block movement of the condition (Gozes O. et al. 2020). Accordingly, compelling monitoring and early conclusion are fundamental for expanding a patient's endurance. Figure 2.1 shows the real-time dataset of COVID-19.

2.1.1 Graph for Deaths

Machine learning (ML) is a process to classify and forecast information from the given dataset. The accuracy depends upon the learning problem. Statistical consistency is a fundamental notation in supervised and unsupervised learning (Jiang F. et al. 2019). Here, we are providing foundations of a unified framework for studying the problem of consistency for a general multiclass learning problem, thereby generalizing many known past results for specific learning problems (Jing Q. et al. 2014). The majority of multiclass learning problems in practice use an evaluation matrix based on a loss matrix, and the most prevalent algorithms for such problems are surrogate minimizing algorithms, which are characterized by surrogate loss. If surrogate loss is convex, then the resulting surrogate minimizing algorithm can be framed as a convex optimization problem and be solved efficiently. The study is divided into three parts: in the first part, we try to describe calibrated surrogate losses which lead to a consistent surrogate-minimizing algorithm for a given loss matrix. We will discuss necessary and sufficient conditions under which calibration happens, based on geometric properties of the surrogate loss and true loss. In the second part, we discuss about convex calibration dimension that characterized the intrinsic difficulty (Lakshmanaprabu S. et al. 2019) of achieving consistency for a training problem. In the last and third section, we discuss generic procedure to conduct convex calibrated surrogate. In the clinical imaging field, PC helped acknowledgment and conclusion (CAD), which truly is a blend of imaging trademark anatomist and ML grouping, has demonstrated the role of planning in supporting radiologists for exact analysis, diminishing the finding time and the expense of determination (Li L. et al. 2020).

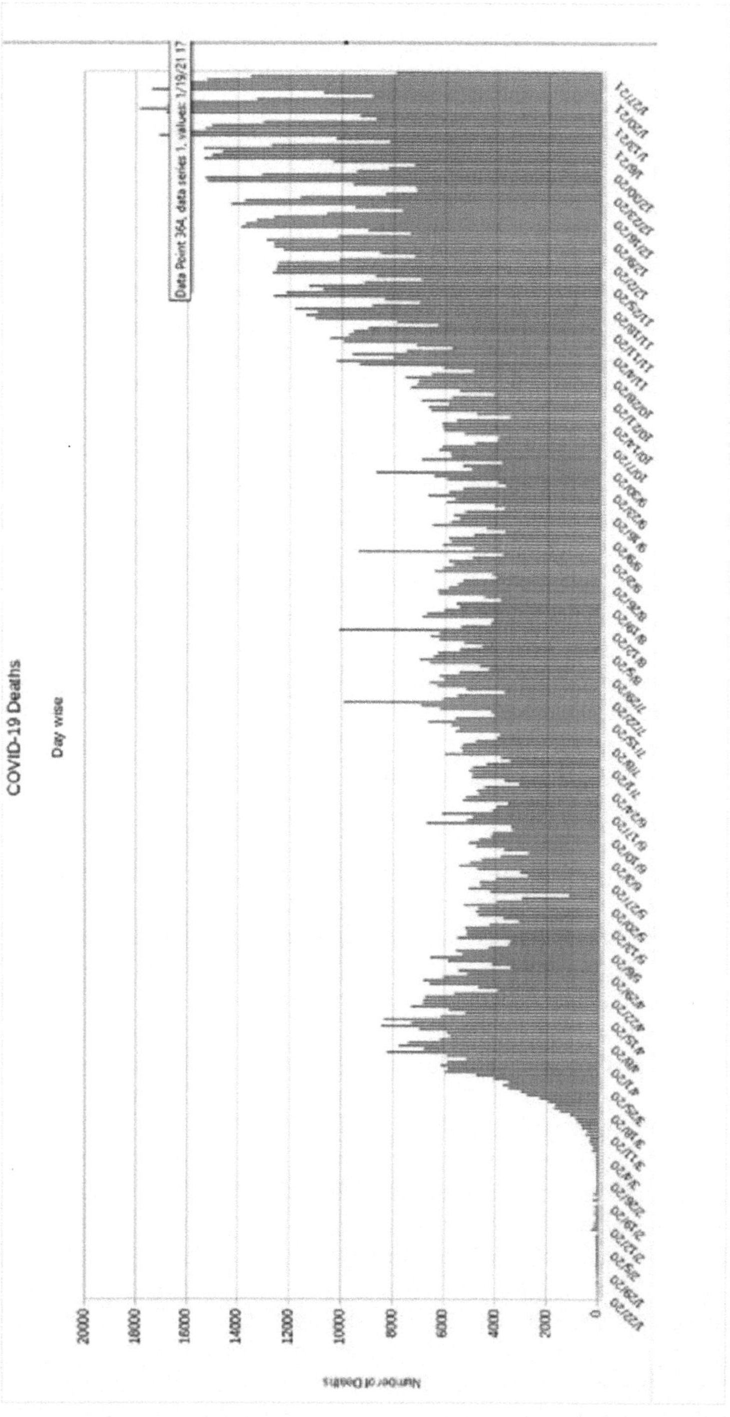

FIGURE 2.1 Real-time dataset of COVID-19.

Source: https://data.humdata.org/dataset/novel-coronavirus-2019-ncov-cases

Deep learning (DL) strategies have indicated promising outcomes in an assortment of PC vision undertakings, for example, division, arrangement, and article discovery. These techniques comprise convolutional layers that can extricate various low-level nearby highlights to be lifted up to a significant level worldwide. A completely associated layer toward the finish of the convolutional neural layers changes over tangled highlights into the probabilities of specific names, for example, clump standardization layer, which standardizes the contribution of a layer with a zero mean and a unit variation, and dropout layer, which is one of regularization strategies that overlook haphazardly chosen hubs, have appeared to improve the exhibition of profound learning-based techniques (Litjens G. et al. 2017). All stuffs are considered to accomplish persuading execution, ideal blends, and structures of the layers just as exact calibrating of hyperparameters is required. This remaining part as one of the principal difficulties of profound learning-based strategies when applied to various fields, for example, clinical imaging (McIntosh K. et al. 2020;Narin A. et al. 2020).

The proposed method in this chapter performs cut-level discovery, also utilizing a lot bigger sample size with better execution analysis. Quiet-level calculations order patients into with and without PCa. It is commonly a moving undertaking to blend ROI-based or cut-level outcomes with quiet-level outcomes. The exhibition of profound learning-based strategies to nonprofound learning put together techniques with respect to the arrangement of COVID-19 virus MRI cuts versus non-COVID-19 virus MRI cuts with 172 patients. They assessed their VGGNet-propelled seven layers (five convolution layers and two internal item layers). CNNs' classifier's presentation is dependent on cross-approval. To start with, they arranged each cut of a given patient and afterward changed over the cut-level outcomes into persistent-level outcomes by a basic democratic technique and accomplished the patient-level AUC of 0.84, affirmative expectation esteem (APV) of 79%, and negative forecast esteem (NPV) of 77% (Pirouz B. et al. 2020). In this work, we accomplished comparable outcomes with an autonomous test set and a bigger sample size. The main characteristics of this project are as follows:

1. In contrast with past endeavors to anticipate PCa repeat utilizing just H&E recolored tissue pictures, our methodology doesn't depend on creating highlights. Rather, we utilize profound figuring out how to naturally get familiar with a chain of command of highlights to differentiate repetitive from nonintermittent morphological examples.
2. We propose a two-phase profound learning-based way to deal with characterized H&E recolored tissue pictures to anticipate a PCa repeat likelihood. This methodology can be expanded to anticipate other fine-grained tissue classes, for example, malignant growth grades, types, and atomic sub-kinds of disease of different organs for exactness treatment arranging. In the first stage, we utilize a convolutional neural system (CNN)to distinguish the atomic focuses with high precision in a given tissue picture. The proposed core location calculation was done to identify both epithelial and stromal cores in COVID-19 just as non-COVID-19 areas of tissue pictures.
3. The subsequent stage utilizes another adjusted CNN which takes patches on the identified atomic focuses as a contribution to assess fixed insightful disease repeat likelihood as its yield. This stage can, without much of a stretch, be modified and retrained to foresee the likelihood of various

sub-kinds of COVID-19 in various sorts of tissue pictures. The final subtype likelihood of a given patient (likelihood of repeat) is controlled by accumulating its fixed shrewd probabilities dictated by the adjusted CNN (Rahimzadeh M. et al. 2020).

4. We use shading standardization as a prehandling venture to fix the impact of undesirable varieties in tissue pictures because of recoloring and examining contrasts across medical clinics. We are considering features of the images based on color combination (Ronneberger O. et al. 2015).

Our pipeline contains five independently prepared modified CNN engineering that is motivated by ResNet, a decision tree-based component extractor, and a Random Forest classifier. For the accuracy of the exhibition, we isolated the dataset into three separate sets, the preparation, approval, and test sets, and guaranteed that the test set was never observed by the classifier during preparing and tweaking. Our classifier's presentation on the free test set was better and more vigorously looked at than comparative examinations that proposed CAD devices for COVID-19 virus discovery utilizing profound modified CNNs (Song F. et al. 2020;Srinivasulu Asadi et al. 2020).

2.2 WHAT WE COVER IN

The corona has four family members Alphacoronavirus, Betacoronavirus, Gammacoronavirus, and Deltacoronavirus. These techniques are applicable to three types of infections excluding Gammacoronavirus. Alphacoronavirus and Betacoronavirus infect only mammals such as humans, pigs, bats, and cats. Gammacoronavirus affects only birds, while Deltacoronavirus infects both mammals and birds. Different family of the virus shows different symptoms in animals such as in chicken, as it causes upper respiratory tract disease in them, but in cows, pigs, and turkeys, it causes bad cold and diarrhea. This is causing with mortality rate approximately 3%—15%. The source of this virus is not yet identified (bats, snakes, etc.).

2.3 LITERATURE SURVEY

Siegel, R.L., Miller, K.D., Jemal file://H:\Asadi Srinivas Books for Preview\Pirouz B, Shafee Haghshenas S, Shafee Haghshenas S, Piro P. Investigating a serious challenge in the sustainable development process: analysis of confirmed cases of COVID-19 (a new type of coronavirus) through a binary classification using AI and regression analysis. Sustainability 2020; 12(6):2427. Pirouz B. et al. (2020) proposed "COVID-19 journal for clinicians." This paper attempts to decipher the real number of deaths caused in 2013 by age owing to driving reasons (12) and owing to COVID-19 disease (5). Population-based malignancy occurrence information in the United States has been gathered by the National COVID-19 Institutes.

The NAACCR announced that all malignancy cases were characterized by the ICD for Oncology with the exception of adolescence and preadult COVID-19s, which were arranged by the ICCC. The reason for death is grouped by the ICD. Whenever conceivable, disease frequency rates introduced in this chapter were balanced for delays in detailing, which happened on account of a slack in the event that catch or information redresses (Srinivasulu Asadi et al. 2020).

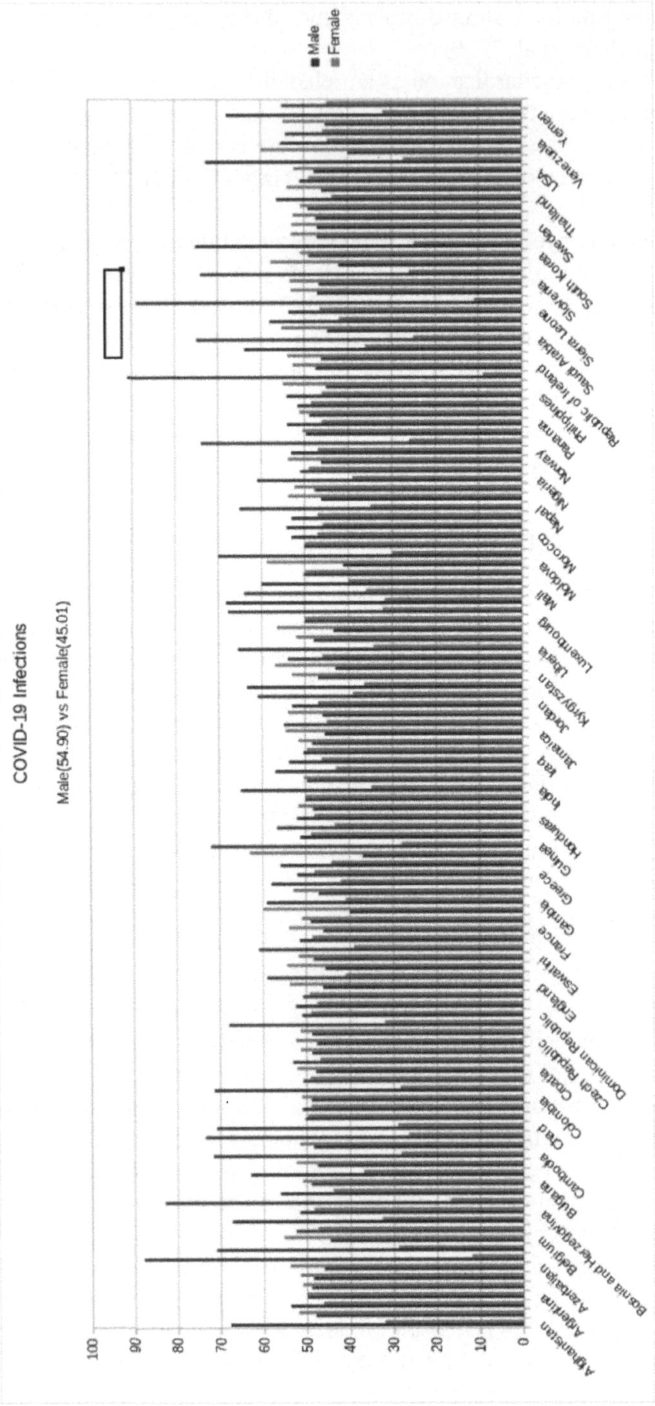

FIGURE 2.2 Male vs. female COVID-19 infection rates.

Source: https://data.humdata.org/dataset/novel-coronavirus-2019-ncov-cases

The latest year for which rate and mortality information are accessible slacks 2 to 4 years behind the present year because of the time required for information assortment, accumulation, quality control, and spread. The quantity of intrusive disease cases was assessed utilizing a three-advanced spatio-worldly model dependent on excellent rate information from 52 states and the District of Columbia representing around 94% populace. This strategy can't appraise quantities of malignancies since information on the event of these COVID-19s is not be required accounted for in disease vaults (Srinivasulu Asadi et al. 2017). Figure 2.2 shows male vs. female COVID-19 infection rates.

The lifetime probability of being determined to have an intrusive malignancy level is more for men (52%) than for women (48%) hazard (Sun D. et al. 2020). Mild and marginal cerebrum COVID-19s are excluded from the 2016 case gauges in the light of the fact that the computation strategy requires recorded information, and these COVID-19s were not required to be accounted for until 2004. COVID-19 frequency rates expanded in kids and young people by 0.6% every year from 1975 through 2012 (Srinivasulu Asadi et al. 2017).

2.4 RELATED WORK

2.4.1 Existing System

Confusion matrix are tables that add to existing network to better detect positive or negative COVID-19 cases and false cases of COVID-19. The output is better overall performance and accuracy. The unstructured COVID-19 image dataset and a few cases of COVID-19, by using the proposed approach, can have better COVID-19 detection along with the other classification detection (Vasilomanolakis E. et al. 2015). The precision of the COVID-19 class is low in findings, in comparing with other detecting COVID-19 from X-ray chest image dataset, the tested neural networks on a larger number of image datasets. The test image datasets were much better than trained dataset images. As is implemented earlier, because of the reason of less number of cases (i.e., 42 cases) of COVID-19 and 12,381 cases from the other two classes, the false positives rate of the COVID-19 class will become higher than true positives rates (Wang L. et al. 2003;Wang S. et al. 2020).

For example, in the first fold, the concatenated network detected 26 cases correctly out of 31 COVID-19 cases, and from 11,271 other cases, it only mistakenly identified 68 cases as COVID-19 (Yang R. et al. 2020;Zreik M. et al. 2018). If we had equal samples from the COVID-19 class as from the other classes, the precision would become high in value. Still, because of having few COVID-19 cases and many other cases for validation, the precision would become low in value.

Disadvantages of Existing System are as follows:

- Less accuracy.
- High time complexity.
- Low performance.
- High computational cost.
- Uses a lot of training data.

2.5 PROPOSED SYSTEM

Modified CNNs are the most mainstream profound learning models for handling multidimensional cluster information —for example, shading pictures. A run-of-the-mill modified CNN comprises different convolutional and pooling layers followed by a couple of completely associated layers to all the while get familiar with an element order and characterize pictures. It utilizes blunder back spread—a proficient type of inclination plunge—to refresh the loads interfacing its contributions to the yields through its multilayered architecture. Figure 2.3 express architecture of modified CNN.

In this chapter, we present a two-phase approach utilizing two separate modified CNNs. The main modified CNN distinguishes cores in a given tissue picture, while the second modified CNN takes patches focused at the identified atomic focuses as a contribution to anticipate the likelihood that the fix has a place with an instance of PCa repeat. Before portraying our modified CNN models, we present the subtleties of the information we used to build up the proposed PCa repeat model.

Advantages of Proposed System:

The advantages of proposed system are as follows:

- High accuracy.
- Low time complexity.
- High performance.
- Reduces the computational cost.
- Even works with a small amount of training data
- Deep convolution neural networks based on the addition of exception and ReNet60V3 to improve the performance and accuracy.
- Proposed training approach for dealing with unbalanced information datasets.

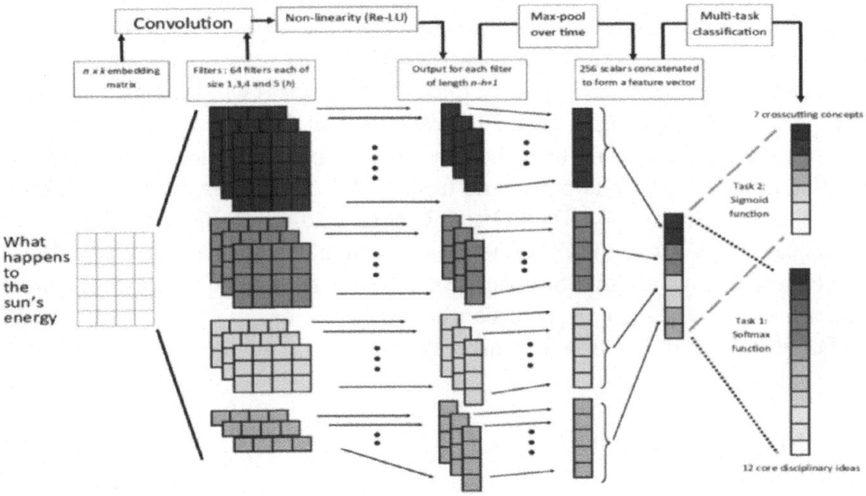

FIGURE 2.3 Express architecture of modified CNN.

- Evaluation of deep networks on 1203 COVID-19 chest X-ray image datasets.
- Evaluated ResNet60V3 and exception on COVID-19 image dataset and compared with a proposed neural network.

2.6 SYSTEM DESIGN AND IMPLEMENTATION

System design is the strategy or forte of portraying the plan, fragments, parts, relations, and information for architecture to fulfill important. It is the utilization of unstructured and structured guesses to think in a right way, with spread and joint effort of sets with structures' assessment, systems building, and plans. Execution or proficiency is calculated as per the yield given by the application. Prerequisite details particularly have a significant influence on the investigation of a framework. Just when the prerequisite details are appropriately given, it is conceivable to structure a superior framework, which will fit into the required condition. It rests to a great extent with the clients of the current framework to give the necessary particulars since they are the individuals who at last utilize the framework.

2.7 PAPER IMPLEMENTATION DETAILS

2.7.1 System Modules

There are three modules:

- Collecting data sources
- Preprocessing datasets
- Feature learning

2.7.1.1 Collecting Data Sources

Here, two types of datasets are used to perform risk assessment.

2.7.1.2 Data in Structured

It cited to information of high level of relation, to such a level that incorporation in a network database is accurate and frequently searched by fundamental, indirect browsing through web computation or hunt activities. This is commonly seen with .csv format.

2.7.1.3 Preprocessing Datasets

- Features, i.e., attributes through which patients are affected are extracted from the datasets.
- Preprocessing may be eliminating duplicate values and adding missing values.
- Each feature importance in affecting the patient can be found using correlation analysis or in max pooling stages.
- In case of unstructured data need to process to structured data with target class

2.7.1.4 Feature Learning

- The extracted features are passed into modified CNN layers to train the neural network structure is used.
- Then extract the high-level features from the modified CNN
- Modified CNN consists of the following layers in this model:

    1. Data input layer
    2. Hidden layers
    3. Data output layer

- This neural network has parameters (w,b) = {h1, h2, h3, b1, b2, b3}.
- The accuracy rate is predicted.
- The results demonstrate that our method has the advantage to infer the prediction of such fatal diseases over the other three methods KNN, decision tree, and Gaussian Naive Bayes).

Here, implementation goes on with structured data and unstructured data, and then the obtained results are compared with performance metrics. Through them, one can know the best method for predicting the risk of hypertension.

2.7.2 Implementing Using Structured Data

This is done using three machine learning (ML) and deep learning algorithms like Bayesian networks, decision tree, and KNN. For implementation, one can use Python, and implementation starts with the following steps,

- Load the dataset in .csv format.
- Then preprocess the data to remove any noisy data.
- Some correlation analysis needs to be done so that how attributes are related to target class using heat map in correlation matrix with diverging palette is known.
- There is a need to divide the dataset for training purposes to build the model.
- Like here, the cardinality of dataset is 303*14. So there is a need to split the data by giving test size and random state. Here, test size is mentioned as 33%; so 203 rows are needed for training the model and the rest 100 rows are needed to test the model, and this could be randomly divided as random state is given.
- Here, while doing this, splitting target class must not be included as the model is developed using only attributes.

Then after splitting, always scale the features after splitting the dataset because we want to ensure that the validation data is isolated. This is because the validation data acts as new and unseen. Any transformation on it will reduce its validity. Now, we are finally ready to create a model and train it. Remember that this is a two-class classification problem. We need to select a classifier, not a regression. Let's analyze three models: decision tree classifier, Gaussian Naive Bayes classifier, and KNN classifier.

2.7.3 K-Nearest Neighbor

The KNN regression is a preparation of dataset and the nearest m case in the preparation informational collection. For KNN, which is required to decide the estimation of separation and the choice of m value? The information is standardized at first in experiment. At that point, the Minkowski distance separation is utilized to gauge the separation. With respect to the choice of parameters m, we found this model is the best fit when m is valued at 6. The algorithm is as follows:

2.7.3.1 Data Input

Information data classification Z, which is a set of training rows and their related labeled classes;

K-value;

2.7.3.2 Output

KNN model ids built

2.7.3.3 Method

1. Start
2. Define k-value and simu() function
3. Train the dataset D
4. For each instance Di in training set and y in test set
5. Evaluate simu(y,Di)
6. k biggest scores of simu(y, Di) has to be found
7. Evaluate simu_avg for KNNs
8. If simu_avg is greater than threshold value, then
9. y is COVID-19 patient
10. otherwise y is not a COVID-19 patient
11. Stop

2.7.3.4 Neural Networks

- Neural Network is a mathematical and computational measure called artificial neural networks neurons.
- Inspired by the neural activity in the brain, they try to replicate similar behavior mathematically (practically far simpler than neurons in the brain.)
- Neural networks are better than traditional ML algorithms and perform well with huge data.
- Because ML algorithm's performance saturates when dataset size grows.
- Input layer is where data is provided.
- Each neuron in the input layer corresponds to the value of a feature.
- Hidden layer is where computation takes place.
- Simple neural networks generally contain one or two hidden layers.
- Output layer generates the result of the neural network.

2.7.3.5 Procedure

Implementing the modified CNN is through using unstructured data because this predicts with high accuracy. The entire computation could be seen in hidden layers. This could be done in five stages.

- At first the unstructured data which consists of patient's data is taken to create vector values.
- Then, the data is organized into two columns such that one with text and the other with target class train the neural network and test the predictions using the vector values.

2.7.3.6 Step 1: Representation of Text Data

1. Each word can be represented as a vector of numerical qualities (a section grid).
2. Analysis is represented as a Zd-dimensional vector, where $d = 60$, i.e., we speak to each word as a section vector (segment framework) containing 60 lines.
3. Presently, the content can be represented by attaching the segment vectors; meaning we can stack up the section vectors one next to the other to make a network of measurement $d \times n$. (Just words are stacked next to each other in a sentence.)

2.7.3.7 Step 2: Convolution Layer of Text MCNN

4. Start a pointer at position 1.
5. Expecting the pointer is at the position I, take words in indexes J-2, J-1, J, J+1, J+2.
6. Transpose every one of them to shape push grids of 50 segments and annex them next to each other, changing them into a solitary column vector of size 50×5.
7. Add the pointer and change to a new line.
8. For first, second, n –1, and nth words, we have openings, i.e., for the essential word, we don't have two past words. In such a case, fill them with zero vectors.
9. Toward the finish of the aforementioned procedure, we get an $n \times 250$ sized network which is our convoluted grid.
10. The weight network W1∈ R100×250 is of size 100×250. This means we are expecting the neural system to separate 100 highlights for us.
11. Presently, we complete the accompanying computation.

$$\mathbf{H^1_{i,j} = f(W_1[i] \cdot s_j + b_1)}$$

12. This is the speck result of lattices. B1 is a segment network of 100 lines. Inclination is utilized to move the learning procedure.
13. Without including it, it is straightforward weighted whole of highlights, and there is no learning procedure.
14. We get a $100 \times n$ element chart h1; f is an actuation work which is utilized to get nonlinearity. We utilized tanh actuation work.

$$\mathbf{H^1 = (h^1_{i,j})_{100 \times n}}$$

2.7.3.8 Step 3: POOL Layer of Text-Modified CNN

15. From the element diagram h1 which is 100 × n dimensional, we pick the most extreme component in each line of the lattice acquiring 100 greatest qualities from each line.
16. From these 100 qualities, we develop a 100 × 1 lattice h2 (segment vector).
17. The explanation of picking max pooling activity is that the job of each word in the content isn't totally equivalent; by most extreme pooling we can pick the components which assume a key job in the content.
18. Before the finish of Step 3, we have separated 100 highlights from unstructured information.

$$\mathbf{H^1{:}h_j^2 = \max_{1<=i<=n} h_{i,j}^1 j=1,2,\cdots,100}$$

2.7.3.9 Step 4: Full Connection Layer of Text-Modified CNN

19. At that point, give this grid as a contribution to a neural system which conveys the accompanying calculation which is like that of in sync 2 (dot result of networks).
20. W3 is the weighted network of full association layer and b3 is inclination.

$$\mathbf{H_3 = W_3 h_2 + b_3}$$

2.7.3.10 Step 5: Modified CNN Classifier

21. Max classifier as yield classifier which predicts the danger of the disease (high or low).
22. This calculation goes with the sigmoid formula and calculates the probabilistic value to predict.
23. The algorithm or step-by-step procedure is as follows:

    # at first, unstructured data is taken and sent to word2Vec algorithm to create vector values.

24. train pd.read_csv(fname)
25. train t rain[TEXT]
26. train train.str.lower()
27. corpus train.str.split()
28. patientcorpus Word2Vec(corpus, size=50,min_count=3)
29. patientcorpus.save(patientwordvec)
30. # Then data is organized so that one contains text and other contains target class.
31. P, Q create_placeholders(n_H, n_W, n_C, n_y)
32. parameters initialize_parameters()
33. R2 forward_propagation(P, parameters)
34. cost compute_cost(R2,Q)
35. Optimizer tf.train.GradientDescentOptimizer(learningRate=learningRate).minimize(costing)
36. inittf.global_variables_initializer()
37. with tf.Session() as sess:

38. sess.run(init)
39. for step in range (numOfSteps):
40. epcost 0.
41. for j in range (0,m):
42. d = j+1
43. temp_cost = sess.run([optimizer,cost]
44. feed_dict={P:P_train[j:d][:][:][:],P:Q_train[j:d][:]}
45. epcost += temp_cost
46. if print_cost == True;
47. print ("Cost after step %i: %f" % (step, epcost))
48. if print_cost == True and epoch % 1 == 0:
49. costs.append(epcost)
50. correct_prediction = tf.equal(R2, Q)
51. predict_op = R2
52. Calculate accuracy using training dataset and test dataset

2.7.4 Logical Flow of Neural Network

In this, one can observe that at first input value is sent to neural network, and then based on weights convoluted matrix is built. Then, predictions are made and are adjusted by calculating squared mean error or loss function. This loss score is done by using gradient descent optimizer, and then weights are updated again and prediction goes on.

2.8 RESULTS

Figure 2.4 projects execution flow of COVID-19 dataset, Figure 2.5 shows execution flow of COVID-19 dataset, and Figure 2.6 shows iteration vs cost.

```
tbarua1@ubuntu:~$ python3 covid.py
2021-01-03 08:15:29.532120: W tensorflow/stream_executor/platform/default/dso_loader.cc:60] could not load dynamic library 'libcudart.so.11.0'; dlerror: libcudart.so.11.0: cannot open shared object file: No such file or directory
2021-01-03 08:15:29.532186: I tensorflow/stream_executor/cuda/cudart_stub.cc:29] Ignore above cudart dlerror if you do not have a GPU set up on your machine.
2021-01-03 08:15:34.647755: I tensorflow/compiler/jit/xla_cpu_device.cc:41] Not creating XLA devices, tf_xla_enable_xla_devices not set
2021-01-03 08:15:34.649432: I tensorflow/stream_executor/platform/default/dso_loader.cc:49] successfully opened dynamic library libcuda.so.1
2021-01-03 08:15:34.728098: E tensorflow/stream_executor/cuda/cuda_driver.cc:328] failed call to culnit: CUDA_ERROR_NO_DEVICE: no CUDA-capable device is detected
2021-01-03 08:15:34.728179: I tensorflow/stream_executor/cuda/cuda_diagnostics.cc:156] kernel driver does not appear to be running on this host (ubuntu): /proc/driver/nvidia/version does not exist
2021-01-03 08:15:34.729063: I tensorflow/compiler/jit/xla_gpu_device.cc:99] Not creating XLA devices, tf_xla_enable_xla_devices not set
Found 13 images belonging to 2 classes. Found 6 images belonging to 2 classes.
2021-01-03 08:15:35.930750: I tensorflow/compiler/mlir/mlir_graph_optimization_pass.cc:116] none of the MLIR optimization passes are enabled (registered 2)
2021-01-03 08:15:35.931657: I tensorflow/core/platform/profile_utils/cpu_utils.cc:112] CPU Frequency: 2194925000 Hz. Epoch 1/5
3/3 [==============================] - 6s 1s/step - loss: 10.6792 - accuracy: 0.5058 - val_loss: 1.3087 - val_accuracy: 0.5000, Epoch 2/5
3/3 [==============================] - 3s 1s/step - loss: 1.3489 - accuracy: 0.3694 - val_loss: 0.6944 - val_accuracy: 0.5000, Epoch 3/5
3/3 [==============================] - 3s 1s/step - loss: 0.6939 - accuracy: 0.4391 - val_loss: 0.7699 - val_accuracy: 0.5000, Epoch 4/5
3/3 [==============================] - 3s 936ms/step - loss: 0.7405 - accuracy: 0.3548 - val_loss: 0.6836 - val_accuracy: 0.6667, Epoch 5/5
3/3 [==============================] - 4s 969ms/step - loss: 0.7814 - accuracy: 0.5058 - val_loss: 0.6822 - val_accuracy: 0.6667,
```

FIGURE 2.4 Execution flow of COVID-19 dataset.

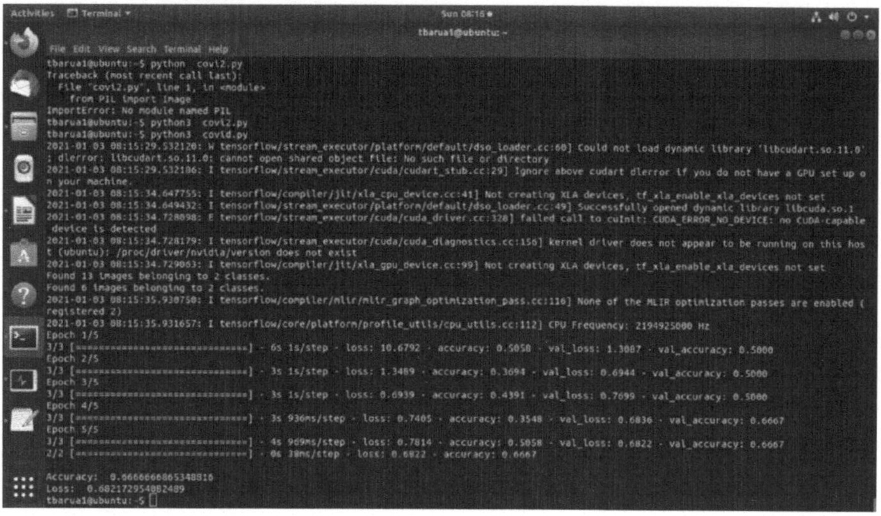

FIGURE 2.5 Execution flow of COVID-19 dataset.

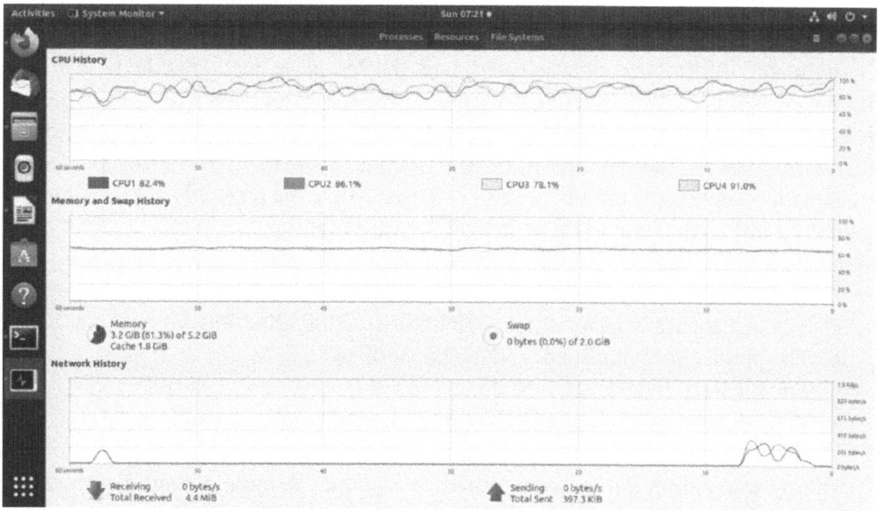

FIGURE 2.6 Iteration vs cost.

For the exhibition assessment in the test, initially, TP, FP, TN, and FN represent obvious positive, bogus positive, genuine negative, and bogus negative, individually. At that point, we can get four estimations: exactness, accuracy, review, and F1-measure as follows:

$$Accuracy = \frac{TP + TN}{TP + FP + TN + FN}$$

$$Precision = \frac{TP}{TP + FP}$$

$$Recall = \frac{TP}{TP + FN}$$

$$F1 - measure = \frac{2XPrecisionXRecall}{Precision + Recall}$$

- The accuracy of decision tree: 71.0%
- The precision of decision tree: 67.24137931034483%
- The recall of decision tree: 79.59183673469387%
- The F1-score of decision tree: 72.89719626168225%
- The accuracy of KNN: 68.0%
- The precision of KNN: 68.08510638297872%
- The recall of KNN: 65.3061224489796%
- The F1-score of KNN: 66.66666666666666%
- The accuracy of Naive Bayes: 84.0%
- The precision of Naive Bayes: 83.6734693877551%
- The recall of Naive Bayes: 83.6734693877551%
- The F1-score of Naive Bayes: 83.6734693877551%

In organized information or structured data, the NB classification is the best in test. In any case, it is likewise seen that we can't precisely anticipate whether the patient is in a high hazard as indicated by the patient's age, sex, clinical lab, and other organized information. In other words, on the grounds that cerebral dead tissue is an ailment with complex side effects, we can't foresee where the patient is in a high hazard gathering just in the light of these straightforward highlights.

- The train accuracy of modified CNN: 1.0
- The test accuracy of modified CNN: 90.9090909090909091%
- The precision of modified CNN: 83.6734693877551%
- The recall of modified CNN: 83.6734693877551%
- The F1-score of modified CNN: 83.6734693877551%

Taking everything into account, for illness chance demonstrating, the precision of hazard expectation relies upon the assorted variety highlight of the medical clinic information, i.e., the advantage is the element portrayal of the ailment, the more the exactness shall be. To find the precision rate, one can arrive at 90.00% to all the more likely assess the hazard. Project Execution Time: It takes few seconds for the execution. The project execution also depends on the system performance. System performance is based on the system software, system hardware, and space available in the system. Figure 2.7 shows COVID-19 input dataset, Figure 2.8 expresses COVID-19 test dataset, Figure 2.9 represents patients with and without COVID-19 virus, Figure 2.10 shows execution time between COVID-19 virus dataset and number of processors, Figure 2.11a shows screenshot of experiment result, and Figure 2.11b projects COVID-19 data size vs. accuracy.

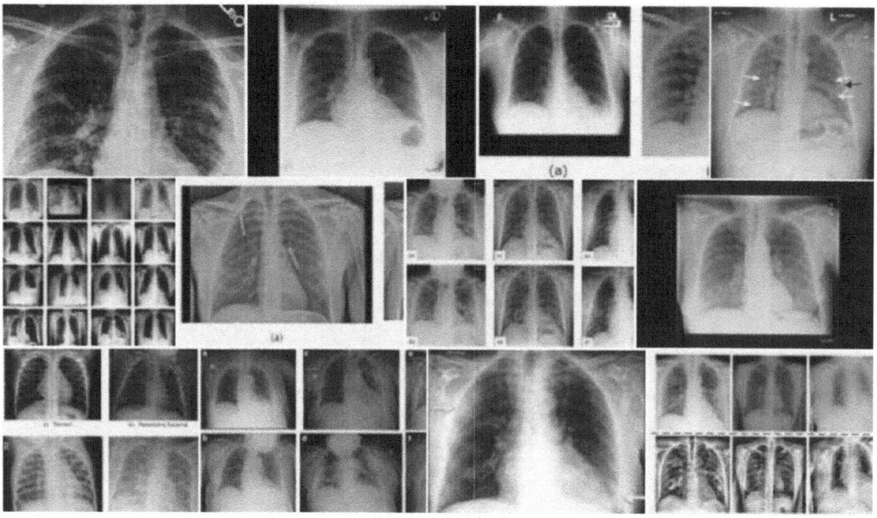

FIGURE 2.7 COVID-19 input dataset.

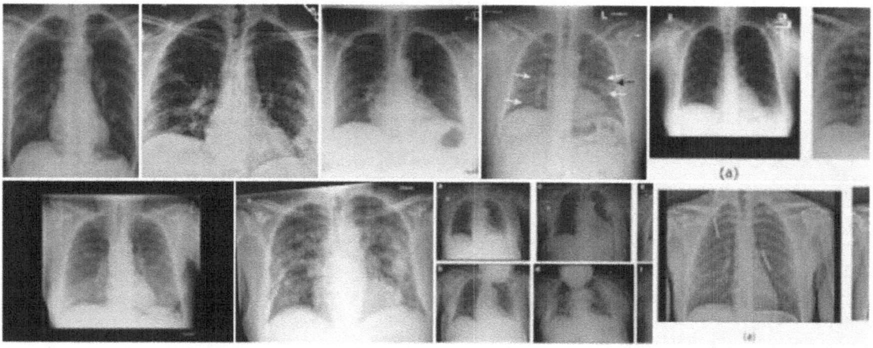

FIGURE 2.8 COVID-19 test dataset.

Input: We are taking 960 image datasets for the input from the database.

Train Data:
Test Data:
Execution Environment:

2.9 CONCLUSION

To the best of the information, none of the current work concentrated on both organized and unorganized information types in the territory of clinical huge information investigation. Contrasted with several typical prediction algorithms, the proposed framework creates high exactness, superiority, and high convergence speed. The

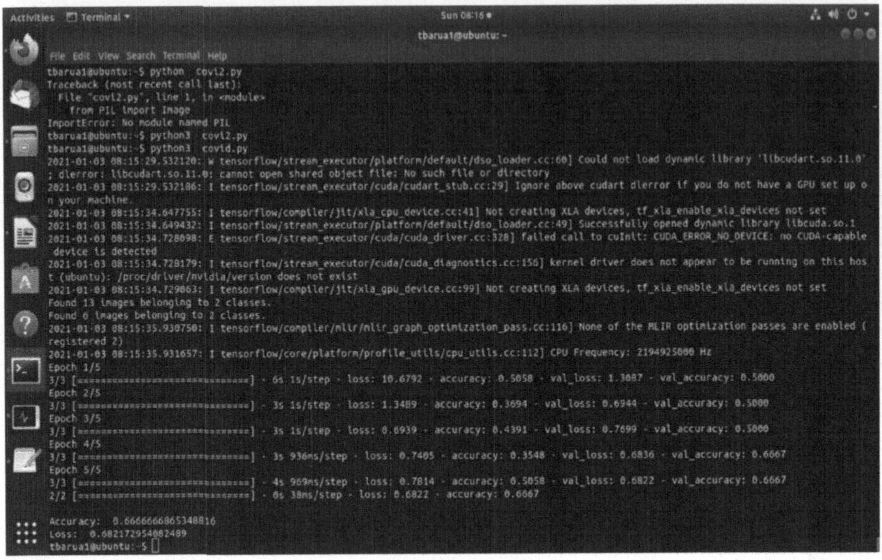

Output:

FIGURE 2.9 Patients with and without COVID-19 virus.

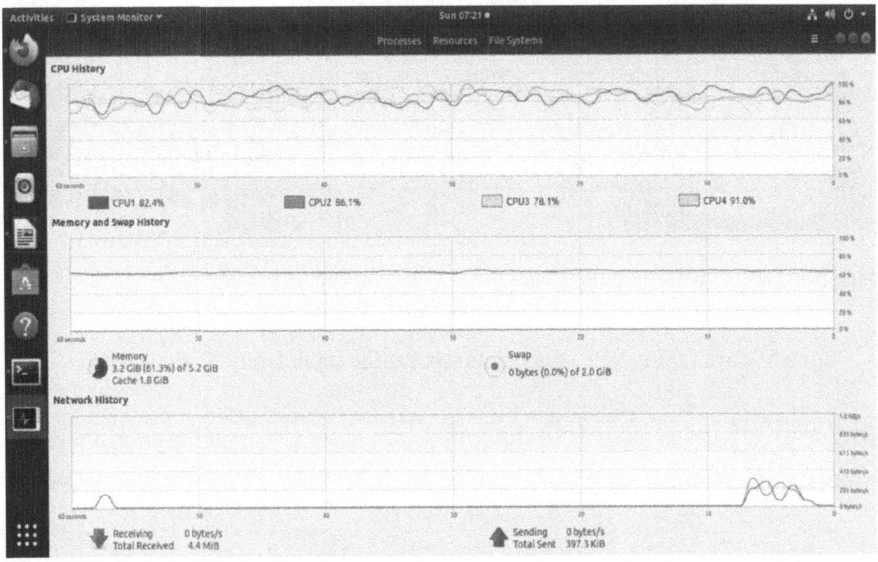

FIGURE 2.10 Execution time between COVID-19 virus dataset and number processors.

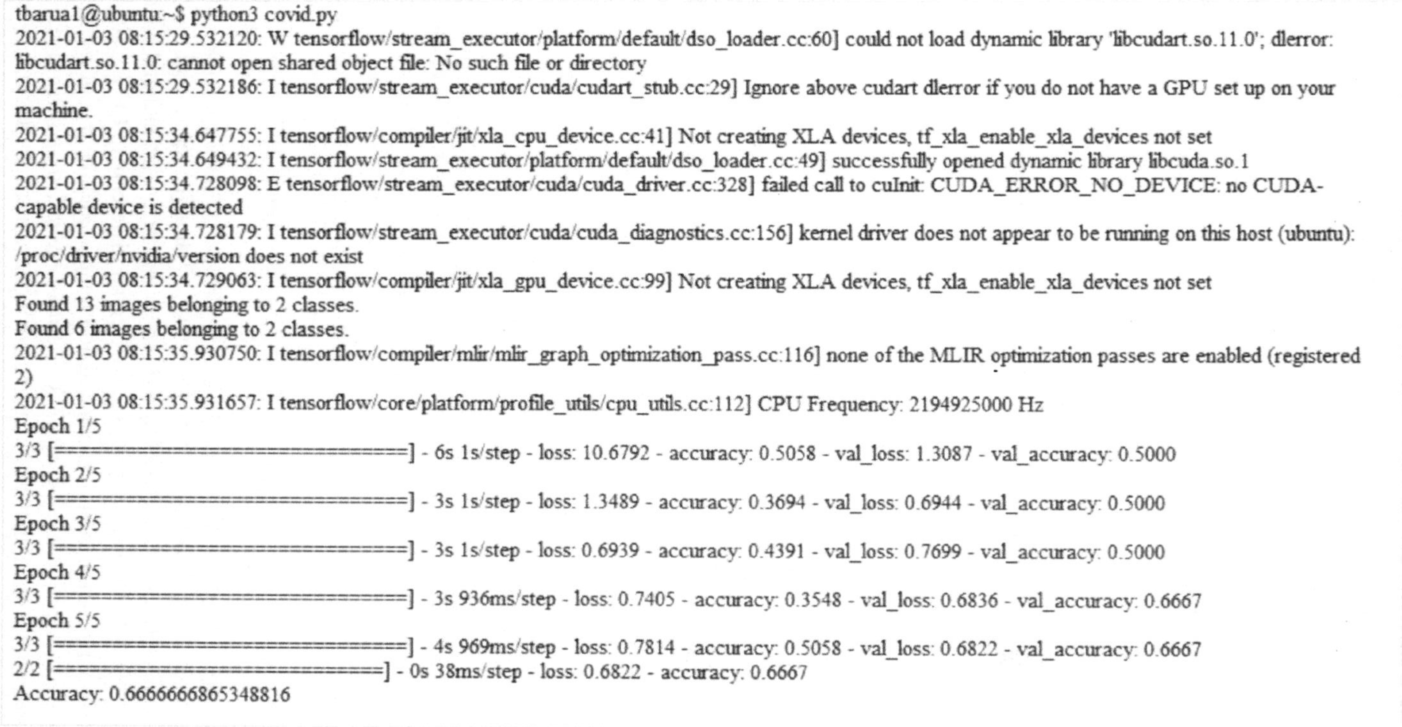

```
tbarua1@ubuntu:~$ python3 covid.py
2021-01-03 08:15:29.532120: W tensorflow/stream_executor/platform/default/dso_loader.cc:60] could not load dynamic library 'libcudart.so.11.0'; dlerror: libcudart.so.11.0: cannot open shared object file: No such file or directory
2021-01-03 08:15:29.532186: I tensorflow/stream_executor/cuda/cudart_stub.cc:29] Ignore above cudart dlerror if you do not have a GPU set up on your machine.
2021-01-03 08:15:34.647755: I tensorflow/compiler/jit/xla_cpu_device.cc:41] Not creating XLA devices, tf_xla_enable_xla_devices not set
2021-01-03 08:15:34.649432: I tensorflow/stream_executor/platform/default/dso_loader.cc:49] successfully opened dynamic library libcuda.so.1
2021-01-03 08:15:34.728098: E tensorflow/stream_executor/cuda/cuda_driver.cc:328] failed call to cuInit: CUDA_ERROR_NO_DEVICE: no CUDA-capable device is detected
2021-01-03 08:15:34.728179: I tensorflow/stream_executor/cuda/cuda_diagnostics.cc:156] kernel driver does not appear to be running on this host (ubuntu): /proc/driver/nvidia/version does not exist
2021-01-03 08:15:34.729063: I tensorflow/compiler/jit/xla_gpu_device.cc:99] Not creating XLA devices, tf_xla_enable_xla_devices not set
Found 13 images belonging to 2 classes.
Found 6 images belonging to 2 classes.
2021-01-03 08:15:35.930750: I tensorflow/compiler/mlir/mlir_graph_optimization_pass.cc:116] none of the MLIR optimization passes are enabled (registered 2)
2021-01-03 08:15:35.931657: I tensorflow/core/platform/profile_utils/cpu_utils.cc:112] CPU Frequency: 2194925000 Hz
Epoch 1/5
3/3 [==============================] - 6s 1s/step - loss: 10.6792 - accuracy: 0.5058 - val_loss: 1.3087 - val_accuracy: 0.5000
Epoch 2/5
3/3 [==============================] - 3s 1s/step - loss: 1.3489 - accuracy: 0.3694 - val_loss: 0.6944 - val_accuracy: 0.5000
Epoch 3/5
3/3 [==============================] - 3s 1s/step - loss: 0.6939 - accuracy: 0.4391 - val_loss: 0.7699 - val_accuracy: 0.5000
Epoch 4/5
3/3 [==============================] - 3s 936ms/step - loss: 0.7405 - accuracy: 0.3548 - val_loss: 0.6836 - val_accuracy: 0.6667
Epoch 5/5
3/3 [==============================] - 4s 969ms/step - loss: 0.7814 - accuracy: 0.5058 - val_loss: 0.6822 - val_accuracy: 0.6667
2/2 [==============================] - 0s 38ms/step - loss: 0.6822 - accuracy: 0.6667
Accuracy: 0.6666666865348816
```

FIGURE 2.11A Screenshot of experiment result.

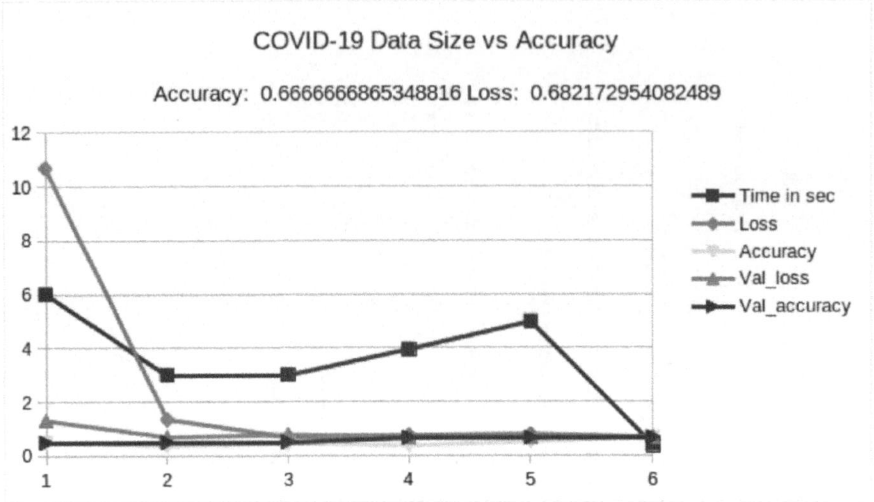

Loss: 0.682172954082489

FIGURE 2.11B COVID-19 data size vs. accuracy.

proposed modified convolutional neural networks algorithm uses organized and unorganized data from hospitals. Data mining ML and deep learning play a vital role in many subjects such as ML, AI, and big data systems. The important objective of this system is to increase the performance and accuracy of the model. This COVID-19 dataset will increase the accuracy of more algorithms, but the CNN and logistic and linear regression perform in a better and improved way. Many advanced approaches will be used to enhance the performance and accuracy of the model in the future. Contrasted with a few regular prediction algorithms, the forecast exactness or accuracy of our proposed algorithm arrives at 91% with an assembly speed which is more accurate than other algorithms.

This chapter presents a combination of neural networks based on CNN and VGGNet networks for dividing the chest COVID-19 images falling into three types of COVID-19 positive, pneumonia positive, and negative cases. Two open-source datasets are the combination of 180 and 960 images' dataset from patients infected with pneumonia and COVID-19. A total of 8,851 images from normal people and a few images of the positive or negative COVID-19 class are presented. The proposed technique for training the convolutional neural network when the dataset is unbalanced. The training dataset is divided into parallel eight phases. There are total 960 images (pneumonia: 434, COVID-19: 249, normal: 277) in each category. The selected number of each class is almost equal to each other in each category so that our network can learn COVID-19 characteristics including the features of the other two classes that are normal and pneumonia. The images from normal and pneumonia classes were different so that the network could differentiate COVID-19 from other classes. The training dataset included 960 images, and the rest of the images were allocated for evaluating the network. We used the tested and tried method to test our model on a large number of images so that our real achieved accuracy would be clear. The proposed model reached an average accuracy of 99.50%, and 80.53% sensitivity, and the final accuracy fell between 92.6% and 1.63%. This model will be publicly available for medical diagnosis and will require larger datasets from COVID-19 positive, and accordingly the accuracy and performance of the proposed model will increase further.

REFERENCES

Botta A, et al. (2016). Integration of cloud computing and internet of things: a survey. *Future Gen. Comput. Syst.*, 56: 684–700.

Cheng J-Z, Ni D, Chou Y-H, Qin J, Tiu C-M, Chang Y-C, Huang C-S, Shen D, Chen C-M (2016). Computer-aided diagnosis with deep learning architecture: applications to breast lesions in US images and pulmonary nodules in CT scans. *Sci. Rep.*, 6(1): 1–13.

Gozes O, Frid-Adar M, Greenspan H, Browning PD, Zhang H, Ji W, Bernheim A, Siegel E (2020). Rapid AI development cycle for the COVID-19 virus (COVID-19) pandemic: initial results for automated detection & patient monitoring using deep learning CT image analysis. *arXiv:2003.05037.*

Jiang F, Deng L, Zhang L, Cai Y, Cheung CW, Xia Z (2020). Review of the clinical characteristics of COVID-19 virus disease 2019 (COVID-19). *J. Gen. Intern. Med.*, 35(5): 1545–1549.

Jing Q, et al. (2014). Security of the internet of things: perspectives and challenges. *Wirel. Netw.*, 20(8): 2481–2501.

Lakshmanaprabu S, Mohanty SN, Shankar K, Arunkumar N, Ramirez G (2019). Optimal deep learning model for classification of lung cancer on ct images. *Future Gen. Comput. Syst.*, 92: 374–382.

Li L, Qin L, Xu Z, Yin Y, Wang X, Kong B, Bai J, Lu Y, Fang Z, Song Q, et al. (2020). Artificial Intelligence distinguishes covid-19 from community acquired pneumonia on chest CT. *Radiology.* https://doi.org/10.1148/radiol.2020200905

Litjens G, Kooi T, Bejnordi BE, Setio AAA, Ciompi F, Ghafoorian M, Van Der Laak JA, Van Ginneken B, Sanchez CI (2017). A survey on deep learning in medical image analysis. *Med. Image Anal.*, 42: 60–88.

McIntosh K (2020). *COVID-19: Epidemicology, Virology, Clinical Features, Diagnosis, and Prevention.* https://www.uptodate.com/contents/covid-19-epidemiology-virology-and-prevention

Narin A, Kaya C, Pamuk Z (2020). Automatic detection of coronavirus disease (COVID-19) using x-ray images and deep convolutional neural networks. *arXiv:2003.10849.*

Pirouz B, Shafee Haghshenas S, Pirouz B, Shafee Haghshenas S, Piro P (2020). Development of an assessment method for investigating the impact of climate and urban parameters in confirmed cases of COVID-19: a new challenge in sustainable development. *Int. J. Environ. Res. Publ. Health*, 17(8): 2801.

Pirouz B, Shafee Haghshenas S, Shafee Haghshenas S, Piro P (2020). Investigating a serious challenge in the sustainable development process: analysis of confirmed cases of COVID-19 (new type of COVID-19 virus) through a binary classication using artificial intelligence and regression analysis. *Sustainability*, 12(6☺: 2427.

Rahimzadeh M, Attar A, et al. (2020). Sperm detection and tracking in phase-contrast microscopy image sequences using deep learning and modified CSR-DCF. *arXiv:2002.04034.*

Ronneberger O, Fischer P, Brox T (2015). U-net: convolutional networks for biomedical image segmentation. In: *International Conference on Medical Image Computing and Computer-assisted Intervention.* Springer, pp. 234–241.

Song F, Shi N, Shan F, Zhang Z, Shen J, Lu H, Ling Y, Jiang Y, Shi Y (2020). Emerging 2019 novel COVID-19 virus (2019-nCoV) pneumonia. *Radiology*, 295(1): 210–217.

Srinivasulu A, Chanakya GM (2017). Health monitoring system using integration of cloud and data mining techniques. *HELIX J.*, 7(5): 2047–2052.

Srinivasulu A, Pushpa P (2020). Disease prediction in big data healthcare using extended convolutional neural network techniques. *Int. J. Adv. Appl. Sci. (IJAAS)*, 9(2): 85–92. ISSN:2252–8814. https://doi.org/10.11591/ijaas.v9.i2.pp85-92.

Srinivasulu A, Rajesh B (2017). Improving the performance of KNN classification algorithms by using Apache Spark. *i-manager's J. Cloud Comput.*, 4(2).

Sun D, Li H, Lu X-X, Xiao H, Ren J, Zhang F-R, Liu Z-S (2020). Clinical features of severe pediatric patients with COVID-19 virus disease 2019 in Wuhan: a single center's observational study. *World J. Pediatr.*, 16(3): 251–259.

Vasilomanolakis E, et al. (2015). On the security and privacy of internet of things architectures and systems. In: *2015 International Workshop on Secure Internet of Things (SIoT)*. IEEE.

Wang L, Wong A (2020). Covid-net: a tailored deep convolutional neural network design for detection of COVID-19 cases from chest radiography images. *arXiv:2003.09871.*

Wang S, Kang B, Ma J, Zeng X, Xiao M, Guo J, Cai M, Yang J, Li Y, Meng X, et al. (2020). A deep learning algorithm using CT images to screen for COVID-19 virus disease (COVID-19). *medRxiv*. https://doi.org/10.1101/2020.02.14.20023028

Yang R, Li X, Liu H, Zhen Y, Zhang X, Xiong Q, Luo Y, Gao C, Zeng W (2020). Chest CT Severity score: an imaging tool for assessing severe COVID-19. *Radiol. Cardiothor. Imag.*, 2(2), e200047.

Zreik M, Lessmann N, van Hamersvelt RW, Wolterink JM, Voskuil M, Viergever MA, Leiner T, Isgum I (2018). Deep learning analysis of the myocardium in coronary CT angiography for identification of patients with functionally significant coronary artery stenosis. *Med. Image. Anal.*, 44: 72–85.

3 Blockchain for Electronic Voting System

Subba Reddy Bonthu, Suchismitaa Chakraverty, Nadimpalli siva subrahmanya Varma, Ramani S, and Marimuthu Karuppiah

CONTENTS

DOI: 10.1201/9781003152392-3

3.1 INTRODUCTION

The dawn of human civilization saw the rise of monarchy across the world. Right from then on, the world has seen many different forms of government including authoritarianism, theocracy, and dictatorship and, finally, evolving into the modern-day democracy. Nonetheless, the immense popularity of democracy lies in the concept of choosing a people's government which, in turn, gives rise to the concept of conducting free and fair elections to choose the people's government. Traditionally, many different forms of voting practices have been in existence. The most notable ones are paper ballots. Though a very simple and cost-friendly method of voting, it suffered the high-end problem of being an insecure method. As a result, the electronic voting system was developed to ensure a traceable and verifiable voting process.

With the evolution of the modern-day distributed computing technology, a highly enhanced tool for recording information has been developed which is termed as the "Blockchain Technology" or "Distributed Ledger Technology." A technology based on the functioning of a peer-to-peer network, it comprises a digital ledger of transactions having the capability of replicating and getting dispersed across all computers on the peer-to-peer network (Adeshina et al. 2019).

The key features that make blockchain technology a strong and undauntable contender for being a viable and secure solution to the E-voting process are its immutability, transparency, digital freedom, decentralized services, better security, cost-effectiveness, and improved efficiency. Thus, being the proud possessor of a multifeathered cap, blockchain technology is, indeed, an ideal contemporary solution to the present insecure, volatile, and vulnerable E-voting system. The remainder of this chapter will describe the development of blockchain technology and its effective deployment in the E-voting system, concluding with a case study of the present-day E-voting system in India—the largest democracy in the world (Abuidris et al. 2019).

3.2 METHODS USED FOR VOTING

As has been discussed earlier, many different forms of voting practices have been in existence over the years. The most popular ones are the paper ballots, E-voting, and i-voting practices which are described in the subsequent sections.

3.2.1 Paper Ballots

This is the most primitive form of voting practice that has been prevalent in society even until the last decade. A very simple procedure to execute, the voter casts the vote for his/her favorite contender through pen and paper. The notable benefits of this method are that it is cost-effective, easy to use, and can be deployed across all sections of society irrespective of the literacy level. No particular high-end technological skill set is required to cast a vote through this method. As a result, for societies having a small and constrained population, paper ballots deem to be the most popular form of voting (Alam et al. 2018). But the problem arises when a society comprises an ever-increasing population, thus giving rise to an exponential increment in the number of voters each time an election is conducted. Serious threats like booth polling and buying of votes undermine the very ideology of a free and fair election and pave the way for a financially powerful and dictator government and not a people's government to come into existence. As a result, due to its many-sided fallacies, this method of voting has eventually been discarded (Adiputra et al. 2018).

3.2.2 E-Voting

A comparatively new voting method, the E-voting or electronic voting practice takes the help of electronic means to cast and record votes. The E-voting practice can be implemented through either standalone systems called the EVMs (Electronic Voting Machines) or computers connected together across the Internet (Al-madani et al. 2020). Specialized vote-polling devices like punched cards and Direct-recording Electronic Voting systems can be used. A plethora of advantages come along with the E-voting practice. These include speeding up of the ballot counting process, reduced expenses on manual counting of votes, rapid publication of election results, remote casting of votes from an individual's location itself, and, therefore, an overall downscaling of the gross election expenses. Though having multiple benefits, the E-voting method suffers from many disadvantages of security issues in terms of electronic frauds and nontransparency of the election operations (Anilkumar et al. 2019). As a result, blockchain technology is introduced in this method to mitigate these fallacies and ensure a strong, free, and fair E-voting system.

3.2.3 I-Voting

I-voting or Internet voting is a specialized form of E-voting system that authorizes polling of votes from any remote location in the world with the help of an Internet-connected computing device (Anjan et al. 2019). This unique solution was first implemented successfully in 2005 by Estonia for its nation-wide elections. Anonymity in vote casting and the ability to undo the already casted votes are the key features of this methodology.

3.3 CURRENT E-VOTING SYSTEM GAPS

Conducting a free and fair election has always been a challenge for any democracy in the world. As the electoral process has evolved over the years from being simple paper ballot voting to more advanced E-voting and I-voting systems, it has brought a plethora of advantages not only for the voters but also for the vote counting and election result declaration authorities. Since, day by day, the current voting system is gradually moving toward a complete Internet-based process wherein votes can be cast remotely using any computer connected to the Internet, and the results for the day can be published immediately after tallying the scores from all voting machines connected to the Internet post the closure of polls (Ayed et al. 2017). But needless to say, this new electoral system has given rise to many serious threats of security, confidentiality, and integrity which are mentioned in the following sections.

3.3.1 Deploying Proprietary Software

The framework of the current E-voting system requires voting software to be installed in the device which will be used for casting and counting the votes polled. In this context, multiple vendors are available today who claim to provide secure and trustworthy software for conducting elections. But, the underlying problem is how far one can trust this proprietary software for conducting a fair and secure election. As a matter of fact, it is not known explicitly that a voting software vendor has not got any implicit partnership with one of the contesting parties which will eventually provide the partner contesting party an undue advantage over the rest of the candidates, thus leading to the conduction of a biased election. To avoid such unscrupulous circumstances, an independent testing authority can be established which will perform thorough audits of the voting software before deploying it for the general elections.

3.3.2 Nontransparency in Enlisting Software Version

The basic objective of conducting a free and fair election is that the entire election process should be transparent. With regard to the current E-voting system, this transparency may include the form of creating laws for mandating the voting software to be kept open for public inspection. Abiding by this law, every vendor should be asked to maintain a site enlisting the past and present versions of its product software. Yet again, the problem of authentication arises whether the version of the software which is being inspected is actually running in the voting machine or not (Bellini et al. 2019). A plausible solution could be to store the source code of the voting software in a secure archive, maintained by a third party before auditing it so that as and when the demand arises, the source code can be released from the archive for inspection after approval from the third party.

3.3.3 Minimal Security against Day-to-Day Attacks

With the widespread growth of Internet services and the increasingly abundant number of devices with Internet access, almost every standalone mainframe system or distributed network suffers the imminent threat from hackers worldwide trying to gain unauthorized access to these systems (Bera et al. 2020). Nonetheless, even after the shielding of independent audits, proprietary voting software is vulnerable to new and unidentified attacks being crafted by innovative hackers day in and day out.

3.3.4 Incompatibility of Voting Machine and Voting Software

Many times, it is noted that if the voting machine and voting software are purchased from two different vendors, they are not compatible with each other. As a result, problems arise in casting and counting votes when the voting software is made to install and run on the voting machine. Therefore, laws should be planned, which will explicitly state the voting standards to be followed by every vendor while developing a voting machine or software so that such incompatibility problems can be mitigated (Bohli et al. 2007).

3.4 INTRODUCTION TO BLOCKCHAIN

A comparatively recent technology, Blockchain or Distributed Ledger Technology, has, indeed, emerged as a quite recent industry trend alike cloud computing, Artificial Intelligence, big data analytics, Internet of Things (IoT), and so on. It is a network-centric information recording technology that has proven to be quite a popular software solution to many budding applications aimed at solving real-world problems (Bulut et al. 2019). The predominant properties of blockchain, which are responsible for its widespread adaptability, are listed in the following.

- A programmable technology
- Decentralized nature of the technology
- An immutable technology
- A secure technology
- A unanimous technology
- An anonymous technology
- A technology based on timestamp
- A technology based on peer-to-peer network
- A technology which provides faster settlement
- A technology based on irreversible hashing algorithms leading to small changes creating wide variations
- A technology which delivers quick response to client requests

(Canessane et al. 2019)

The utility of blockchain technology lies in the features listed in the following:

- It experiences less failure as compared to other network-based technologies
- It provides complete control to users with its decentralized nature
- It provides tough resistance to malicious attacks and, therefore, quite secure against breakdown
- Since the technology is based on algorithms, it is scam-free, transparent, and authentic

(Casado-Vara et al. 2018)

Since a blockchain's working is based on a peer-to-peer network, the core components which comprise its architecture are:

- *Node*: It is defined as a client device in the blockchain network
- *Transaction*: It is defined as the most basic unit of the blockchain network
- *Block*: It is defined as the data structure which stores the transaction set, and it is disseminated across all network nodes
- *Chain*: It defines the order in which the block is arranged in the network
- *Miners*: These are nodes that are responsible for validating the blocks
- *Consensus*: It defines the set of guidelines or rules used for performing the blockchain operations

(Chafe et al. 2021)

Figure 3.1 shows blocks in a blockchain containing the data part, Hash ID, and Previous Hash ID.

The block of a blockchain network comprises three parts which are listed in the following:

- Information part
- Hash/Unique Block ID
- Previous Hash

(Chaudhari et al. 2018)

FIGURE 3.1 Blocks in a blockchain containing the data part, Hash ID, and Previous Hash ID.

Source: geeksforgeeks.org

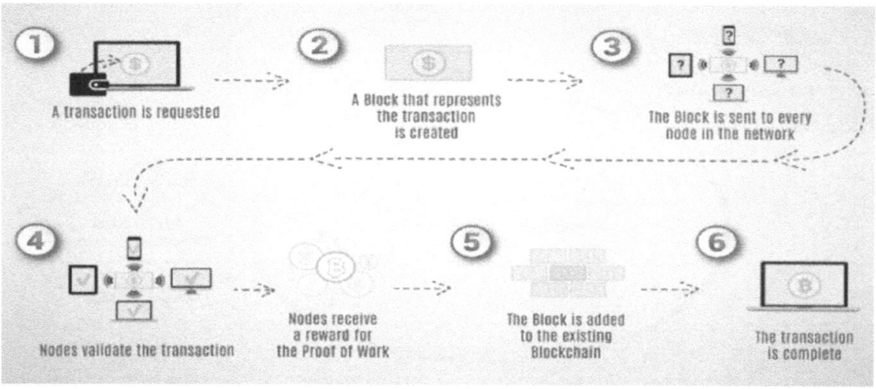

FIGURE 3.2 Working of a blockchain.

Source: specbee.com

Figure 3.2 shows the working of a blockchain. A blockchain works on the following four contiguous steps of operation as listed in the following:

- A transaction record is created.
- Each transaction is validated independently for its legitimacy by all nodes of the blockchain network.
- After validation, the legitimate transaction is appended to a hashed block.
- At the end, the block is appended to the back end of blockchain

(Chaum et al. 2005)

3.4.1 Blockchain Network

At present, there are four types of blockchain network available which are listed in the following.

- *Public blockchain*: It is an open blockchain network that is available for usage by any individual with access to the Internet. It is primarily used for mining and cryptocurrency exchange. It can be either secure or unsecure depending on the stringency adopted by the users in following the security protocols of the network. The well-known examples are Bitcoin and Litecoin. Figure 3.3 shows high-level architecture of public blockchain.

- *Private blockchain*: It usually operates in a closed network and requires permission to get access. Normally, a controlling authority is in charge of the setup and maintenance of a private blockchain network. It is used primarily for voting purposes and supply chain management. The well-known examples are fabric and sawtooth (Cruz et al. 2017; Dalia et al. 2012). Figure 3.4 shows high-level architecture of private blockchain.

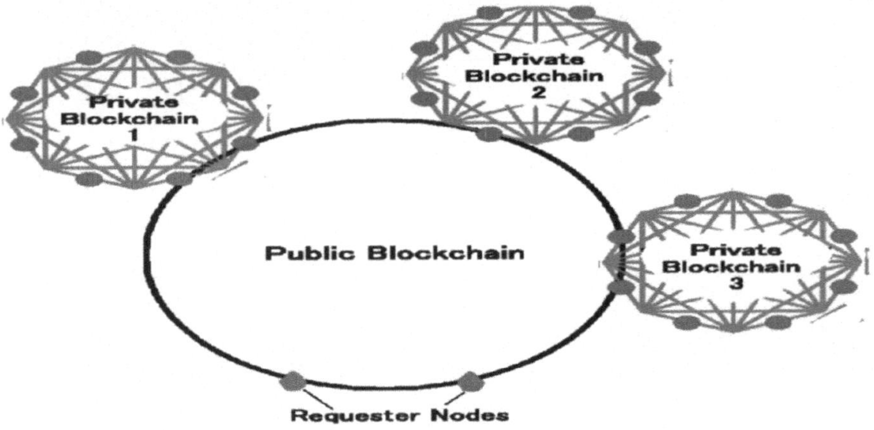

FIGURE 3.3 High-level architecture of public blockchain.

Source: researchgate.net

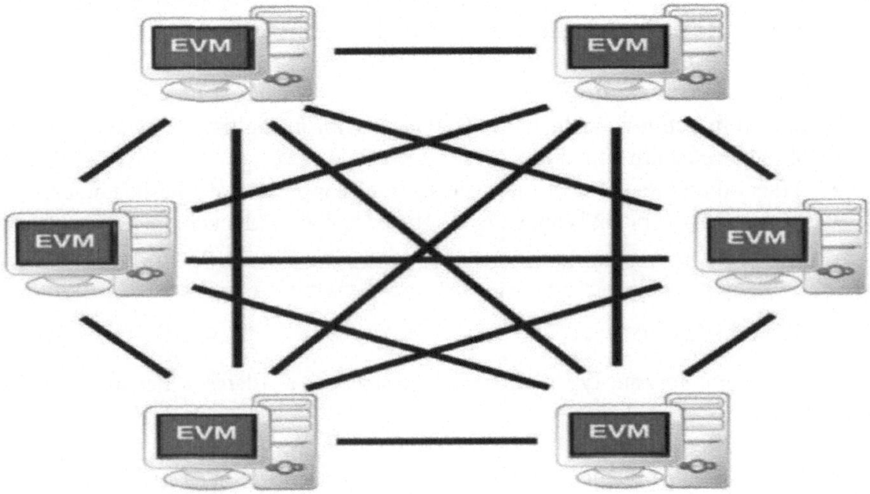

FIGURE 3.4 High-level architecture of private blockchain.

Source: hackernoon.com

- *Consortium blockchain*: It is a partially decentralized blockchain network that is deployed and managed by multiple organizations unlike a private blockchain with a single-control authority. More than one organization can behave like nodes in this blockchain. It is also used for information exchange and mining like public blockchain and commonly deployed by banks and government agencies. The well-known examples are R3 and Energy Web Foundation (Garg et al. 2019). Figure 3.5 shows high-level architecture of consortium blockchain.
- *Hybrid blockchain*: It is an amalgamation of public and private blockchain networks. It allows users to control access rights to selected resources in the network at one's own will. It is flexible and comparatively more secure since permitted blocks can undergo double hashing both from private and public networks. A well-known example is Dragonchain (Han et al. 2020). Figure 3.6 represents high-level architecture of hybrid blockchain.

3.4.2 Countries that Used Blockchain for Voting

General elections have several complications such as shortage of translucency, hacking of political parties' important data, changing of votes, accessibility of votes, and also safety of votes.

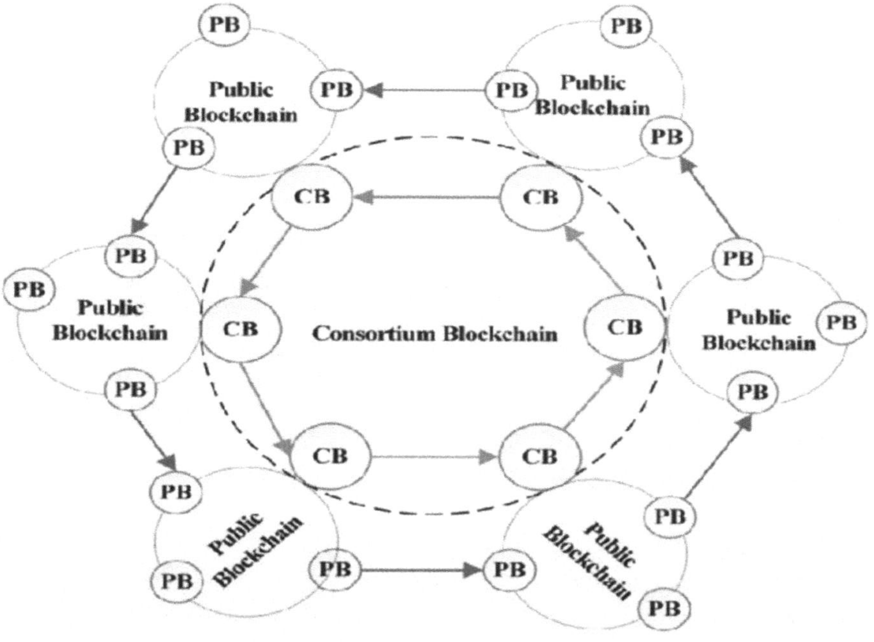

FIGURE 3.5 High-level architecture of consortium blockchain.

Source: Consortium Blockchain-based Malware Detection in Mobile Devices, IEEE Access, Volume 6, 2018

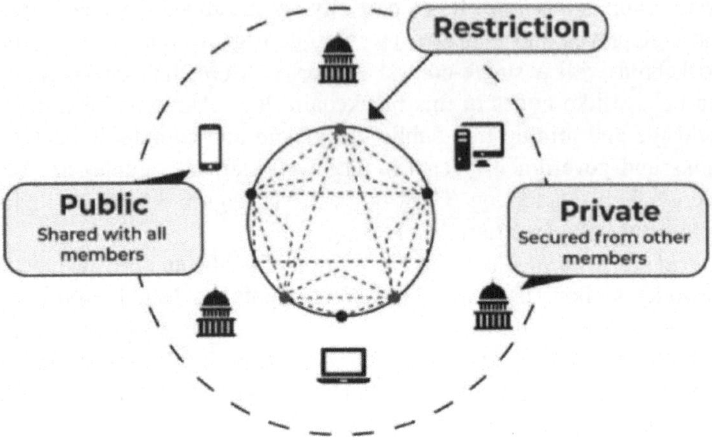

FIGURE 3.6 High-level architecture of hybrid blockchain.

Source: in.pinterest.com

These days, general elections are encountering many more problems. Accordingly, the main problem in general elections is corruption. This may lead to an environment where the people lose faith in the system as well as it creates misunderstandings among people. Consequently, nowadays most of the nations are adapting blockchain for creating a system that can be fastening the work without affecting the regulation of the system and intensify the translucency (Hanifatunnisa et al. 2017; Hao et al. 2010).

3.4.2.1 Sierra Leone

On March 7th, Sierra Leone led the blockchain-based voting system, and it is said to be the first and only nation to become so. Leonardo Grammar, who belongs to Agora, stored the votes in a scattered record, which are unchangeable; therefore, blockchain instantly certifies election's outcome correctly.

Sierra Leone created an environment for the voters with faith and translucency in a controversial election. To document ballots and outcomes as a source, it utilized blockchain, through which lawfulness in the election was maintained and reduced the issues with the opposition parties (Hjálmarsson et al. 2018).

3.4.2.2 Russia

In 2019 June, a blockchain-based E-voting system was organized by the Officials of Moscow. For this project, they collaborated with the election commission of Moscow city and Moscow's department of information and technology, and they accomplished the project successfully. For testing the project, they supplied the needed support. For the normal voting system, this technology based on blockchain is not considered as a substitution, but for the Muscovites, it is used as a different form of voting (Jafar et al. 2020; Karuppiah et al. 2014). The consequences of the electronic voting are stored by the Russian Federation by using Distributed Ledger Technology. This aided in intensifying the translucency of elections and eliminating arbiters in the election procedure (Karuppiah et al. 2015; Kirillov et al. 2019; Lin et al. 2019).

3.5 WORKING OF E-VOTING USING BLOCKCHAIN

The activities that are first attached to the block are the activities that represent the contenders peculiarly. The records when generated will contain contenders' first and last names, and it will be considered as the base block, with each and every contender vote to that particular contender placed at first. Contrary to the remaining transaction, for calculating a vote the foundation will not be counted, and in it, there will be only the contender's name. A protest vote will be allowed by our electronic voting system, where a contender may give an empty vote for revealing the disquiet with every contender or declaring a refusal for the present political election or system (Karuppiah et al. 2019; Khan et al. 2018).

The blockchain will be updated every time the contender votes the records. The block will be having the previously voted voter's data in order to make sure that the system is updated. If any one of the blocks were imperiled, since all the blocks are connected to one another, it will be much easier to search for (Kshetri et al. 2018; Kumar et al. 2019). It is not possible to corrupt as the blockchain is spread out, and not even a single failure example is found. The actual voting takes place at the blockchain. The node which is present in the system any one of them will get the contenders vote and that vote will be attached to the blockchain by the node. To make sure that the system is unified, the voting processor will be having a node in each and every district (Liefhebber et al. 2017).

3.5.1 Requesting for Vote

Contender should use his own credentials to log in to the voting system; the electronic voting system will be utilizing his/her Social Security Number, their residential details, and their confirmation number for the voting given to the registered voter by their respective authorities. Before the contender casts his vote, the vote will be validated, and the system will verify the information if the data matches; then, only the contender will be able to cast his particular vote. The voting system does not permit them to create their identities on their own by the contenders. Arbitrarily generated identities that are generally at risk are systems that permit the Sybil attackers where the attack will be declared in a huge number of duplicate identities, and they will fill the ballot boxes with their illegal votes (Liu et al. 2020; Mitra et al. 2020).

3.5.2 Casting the Vote

Contenders can vote one of the candidates, or they can cast a protesting vote on their wish. Casting the vote by the contender will be done in a very understandable user interface. To every contender, there will be a ticket provided like the Ethereum, with the beginning Boolean expression 1, which, once the vote is cast, will become 0.

A contender will be able to cast a vote only if the Ethereum is one. Revolting issues will be resolved in this manner (Olaniyi et al. 2016; Pandey et al. 2019).

3.5.3 Encrypting Votes

The system will generate an input after the contender casts their respective vote and that has the voters' verification ID followed by their full details of the contender and also the hash value of the last vote. In this manner, each single input can be varying, and this makes sure that the output which is encrypted will also be different. In each block header, the encrypted data will be kept of each and every vote casted. Using SHA, the data relating to every vote and the voter's details will also be encrypted by the single-way hash method that will have no way the data can be retrieved from the encrypted data (Patidar et al. 2019; Patil et al. 2018). The convenient method to reverse the hash could be of guessing encryption way and pit information and then hash to watch whether the results match or not. This manner of hashing makes the votes clearly difficult for reverse engineering techniques such that there is no way that the voter's data would be recovered (Pawade et al. 2020).

3.5.4 Appending the Vote

The data that is stored in the respective blockchain will be available only after creation of a block and subjected to the candidates picked. Each block will be linked to its last casted vote (Pawlak et al. 2019).

3.6 BLOCKCHAIN AS A SERVICE

These are the services provided by the blockchain network to the organizations to build their blockchain applications. It gives the organizations to leverage the cloud-based solutions to develop, operate, and host their own applications (Poniszewska-Marańda et al. 2020).

3.6.1 Smart Contracts

Smart contracts taken out in localized surroundings are responsible and also irreparable applications. No one can change or edit the code or its implementation once the smart contract has been utilized. As discussed, smart contracts make sure to stick parties together to a conclusion. A well-formed trust is built, and the relationship does not rely on a single party. As they are self-verifying and self-executing, smart contracts permit better management for understanding and administering digital accordance (Punithavathi et al. 2019; Roh et al. 2020).

3.6.2 Noninteractive Zero Knowledge Proof

Zero-knowledge verification is another theory that is not directly connected to the blockchain, but it can be considered as a very important aspect for fulfilling some demands to establish an electronic voting system on a blockchain. Cryptographical way by which in a single party, the attester, is able to attest to a second party, the authenticator that proves the value of x, without telling statistics further than that verifier knows the value of x is a zero-knowledge proof (Sadia et al. 2020).

3.7 BLOCKCHAIN AS A SERVICE FOR E-VOTING

Blockchain based and non-blockchain based, and gauges their respective possibility for running a national electronic voting system are the existing electronic voting systems. We approached with a blockchain form electronic voting system which will enhance the specifications that are identified. In the next subsection, we start by identifying the roles and components for executing an electronic voting smart agreement, and, then, we gauge various blockchain frameworks that can be utilized to grasp and establish the election smart contracts.

3.7.1 Election as a Smart Contract

In the agreement procedure, defining a smart contract includes recognizing the roles that are involved in the agreement and the different components and transactions. We start explaining the election roles.

3.7.2 Election Roles

3.7.2.1 Election administrators

It organizes the process of an election. In this role, multiple trusted institutions and companies are signed up. The election administrator members specify which type of election it is, and they launch the forgoing election, arrange the ballots, register the voters, and decide the timeline of the election and assign permissions nodes.

3.7.2.2 Voters

The voters can verify for elections to which they are eligible to vote for and load election ballots, cast their vote, and validate their vote after an election is completed. Voters will be rewarded for voting with tokens when they will cast their vote in an election in the future, and this could be integrated with a smart project.

3.7.2.3 District nodes

Each ballot smart contract representing each one voting district is deployed onto the blockchain when the election administrators or organizers start an election. District nodes are allowed to interact with their respective smart ballot contract when the ballot smart contracts are generated. When an individual voter casts his/her vote from his corresponding smart contract, the district nodes will check the vote data, and every vote they agree on is included onto the blockchain when block time has been reached.

3.7.2.4 Bootnodes

Each institution, with authorized access to the network, hosts a booth node. A booth node helps the district nodes to find each other and communicate. The booth nodes do not keep any state of the blockchain and are carried on a static IP so that district nodes find their peers quicker.

3.7.3 Election Process

A group of smart contracts represent each election procedure, which are shown by the election administrators on the blockchain. Several smart contracts will be elaborated in an election under smart contract for each and every voting district of the election. For each and every voter with its respective voting district location, defined in the voter's registration phase, the smart contract with the communication location will be reminded of the contender after the user verifies himself when voting.

3.7.3.1 Election Creation

Election administrators use a localized application to generate election ballots. This localized app interconnects with an election creator smart contract, a list of contenders, and voting districts in which the admin defines. A set of ballot smart contracts launches them onto the blockchain, with a list of the contenders, for each voting district, where each voting district is a work in each ballot smart contract by a set of smart contractors. When the election is started, permission will be given to each of the belonging district nodes to interact with its belonging ballot smart contract.

3.7.3.2 Voter Registration

Election admins sort the registration of contender's phase. Election admins must define a permanent list of eligible contenders when an election is started. For verification services to securely audit and permit eligible ones, the government requires an element. Using such audification services, each of the allowed voters must after the user verifies him when voting. For each permitted contender, a belonging wallet would be created for the contender. For each and every election, the wallet created for the contender should be varying, and to create such other wallets they could be combined so that the system does not identify itself, which wallet matches a single contender.

3.7.3.3 Vote Transaction

When a contender casts his vote at a voting district, the contender interconnects with a ballot smart contract with the same voting district as defined for any one single contender. The blockchain is interconnected with the smart contract via the communicate district node, which attaches the vote to the block chain if an agreement is received among the majority of the belonging district nodes. Each and every vote is set aside as an exchange on the blockchain, whereas each contender should forgo the transaction ID for their vote for validating purposes. Blockchain containing each transaction clutch data about who voted, and the geographical position of the vote discussed earlier. Belonging ballot smart contract, each vote is joined onto the blockchain if and only if every belonging district nodes accepts on voter data verification (Shahandashti et al. 2016).

3.7.3.4 Tallying the Results

The tallying of the election results is carried on the fly in the smart contracts. Each ballot smart contract for the corresponding location carries its own tally in its very own storage. The final result for each smart contract is produced when an election is completed.

3.7.3.5 Verifying the Vote

Each and every voter receives the transaction ID of his particular vote, as discussed earlier. After verifying personally, using his particular electronic ID and its respective PIN, each individual contender can visit his government official and present their transaction ID. The government official, accessing district node access to the blockchain, uses the blockchain explorer to discover the transaction with the corresponding blockchain using the transaction ID. The voter will be able to watch his particular vote on the blockchain by authenticating that it was carried out correctly (Singh et al. 2018).

3.7.4 Evaluating Blockchain as a Service for E-Voting

For executing and establishing the smart contracts-based elections, there are three different blockchains. They are Exonum, Geth, and the quorum.

3.7.4.1 Exonum

This blockchain uses a programming language called Rust; the Exonum is well built from its start to the end until the execution is completed fully. For privatized blockchains, Exonum is built. This blockchain has a default algorithm known as the Byzantine which is used for conquering the network. With the usage of the default algorithm in the blockchain, there is a chance of around 5,000–6,000 transactions per second which can be handled using the Exonum. Woefully, there is currently only one language that is used for programming in the Exonum which is the main curb in the framework, which the initiators are limited to constraints that are accessible from the language. To overcome these issues, Exonum is trying to focus on these issues and trying to produce a platform that is independent of the description of the interface which makes sure the Exonum will be more user-friendly in the upcoming years.

3.7.4.2 Quorum

Quorum is with transaction privacy and fresh consensus mechanisms. It is an Ethereum-based distributed ledger protocol. It is modernized in line with Geth releases, and it is a Geth fork. Quorum switched up to the consensus mechanism and completely focused on consortium chain-related consensus algorithms. Using this consensus, it permits to keep out more than a hundred transactions per second.

3.7.4.3 Geth

It moves smart contract-based applications just as it is programmed without risk of the downtime, censorship, and fraud interferences, and it is among the three genuine executions of Ethereum protocol. The growth of the Geth protocol is maintained by this framework and is the most developed as well as user-friendly protocol frameworks we have evaluated. The total transactions per second depend on whether or not the blockchain is executed as a public or private network. To base our work, we have chosen Geth as a framework; for such systems, any blockchain framework with the same abilities as Geth must be considered (Suralkar et al. 2019).

3.8 CURRENT PROPOSED SOLUTIONS IN E-VOTING SYSTEM

As technology is improving day by day, the solutions that are being built for the problems using modern technology are also improving. So, in the last few years, there are various proposed methodologies and solutions for the voting system that is taking place currently for replacing it with an E-voting system using various available technologies in the market such as the blockchain.

Some of the proposed solutions for the E-voting system are:

1. General
2. Coin based
3. Integrity of data
4. Consensus

3.8.1 General

There had been many conclusions made by the researchers and by pointing out that in the blockchain platform we can make use of the smart contracts which can help in the significant growth in the E-voting system. It permits cost efficiency, and it provides various possibilities to overcome the barriers in the current E-voting system as well. After comparing both the distributed systems and the present E-voting system, it can be concluded that the governments can make use of the blockchain technology in terms of gaining the voters' confidence and their accountability (Taş et al. 2020).

The blockchain technologies can be considered as the best solution for supporting some of the complex applications like electronic voting systems with security, transparency, and reliability. In the earliest days of the voting system, the people who paid the money were authorized to cast their respective vote. Later, it had been taken into consideration that the voters can cast their vote only once. And, there has been a proposal of various systems so that the voters get a chance to update or change their vote before the election process ends which does not offer any privacy, auditability, and consistency.

In recent years due to the increase in the usage of cloud services, the blockchain-based E-voting system has been developed in the cloud-based environment. The services provided by the cloud give many advantages like simple access to the services, an increase in the performance, and decreased costs as well. But with an improvement in the technology, there were also some user-related problems and technical issues such as chances of occurring delayed transactions if there has been an increase in the load on the main node of the blockchain server. There are other issues regarding the blockchain-based E-voting system such as the lack of the user's knowledge regarding the technologies and scalability of the system (Vo-Cao-Thuy et al. 2019).

3.8.2 Coin Based

In today's world, there has been a spike in the usage of cryptocurrencies in various fields where their key function is to provide privacy and security to the user, and hence they can be used for the blockchain-based E-voting system as well. One of the cryptocurrencies is the Ethereum which can be used for E-voting where it runs on an open platform. Similarly, there are various methods available using different cryptocurrencies. These coin-based solutions provide security, privacy, as well as integrity to the user.

3.8.3 Integrity of the Data

It is the most crucial part that needs to be considered because all the data regarding the voters voting are stored in the database. This includes factors like data transfer securely, storing the data securely, and also to decide whether the user is the right voter or not. However, ascertaining the integrity of the data and as well as the security is more difficult. So, blockchain technologies can be used in the databases to prevent the problem of data manipulation.

In the voting case scenario, there are two approaches for authenticating a right user. One way is by using their government ID card and mobile number during the registration phase along with a password. The other way of authentication is by using biometric information of the user such as face recognition and fingerprint scanning. Therefore, by using these authentication methods, the issue of the duplication of votes can be eliminated by removing the users with the wrong credentials and as well as the user with fake verification. With this process and by integrating the data, the voting process will be maintained in a much secure, transparent way and with much more accurate results.

3.8.4 Consensus

Since the blockchain technologies are a decentralized network that provides transparency, privacy, and security during the transactions. Every transaction that is happening in the network is considered as a verified and secured transaction even though there is no central authority present. This is attainable only because of the consensus mechanism that is being present to check the transactions in the blockchain network which is present in the core part of the network. It is a process that all the nodes in the network come to a final agreement regarding the current scenario in the network. This is the reason for establishing trust between the unknown nodes in the network (Wu et al. 2018).

There are two different types of consensus algorithms:

1. Competitive
2. Noncompetitive [33]

3.8.5 Competitive Consensus

These algorithms ensure the integrity of a blockchain network by taking only a single consensus which assures the state of their neighboring consensus at many sites in an untrusted network system at a similar time. In such algorithms, the complete nodes that are included in the network can also be included in the negotiation even though the nodes are not directly taking part in the participation. Still, there might be a possibility of an event occurring twice in these algorithms because the branch arises when the nodes that are competing produce the blocks simultaneously.

There are different types of competitive consensus algorithms:

1. Proof of work
2. Proof of stake
3. Delegated proof-of-stake

3.8.6 Proof of Work

The proof of work (POW) ensures that the transactions are verified as distinctive and trustworthy as well. The POW is used for implementing a major component which is the key for maintaining the network layer secure and for generating the blocks. The POW had provided the payment system that is decentralized on a peer-to-peer network without requiring any intermediates between the transactions and also decreasing the complete cost. The example of a POW system is Bitcoin.

The POW makes use of a hardware known as application-specific integrated circuits which helps in solving the complex cryptographic algorithms and problems. But the POW is expensive because it needs a lot of energy, and the hardware that POW requires is very expensive. Yet, the POW is very secure blockchain and is more reliable.

3.8.7 Proof of Stake

Proof of stake (POS) is also a possible algorithm and can be considered as a substitute for the POW. These systems are mainly developed to solve the problems that are arising in the blockchain systems that are based on the POW. POS mainly focuses on the energy usage and the hardware. The blockchain in these systems is secured in a very deterministic way. While these systems mainly depend on the external factors such as the hardware and energy consumption, the blockchains in these systems are secured by the internal factors like the cryptocurrency. The example of a POS system is Ethereum cryptocurrency.

It is difficult for any attacker to attack any of the POS systems, as the attacker needs to own a minimum amount of 51% to make the attack successful which makes the attacker to pay for higher costs. But if the attack which the attacker is trying gets failed, it may lead to a huge loss for the attacker. So, attacking a POS system is quite a difficult task for the attackers. Figure 3.7 shows working of a POS system.

FIGURE 3.7 Working of a POS system.

Source: Ledger

3.8.8 Delegated Proof of Stake

This algorithm is one of the variants of the POS consensus. In delegated proof-of-stake (DPOS), the blocks are signed by the designated representatives. The one who has the largest balance in the network chooses their corresponding representatives, and each of them gets the right to sign the blocks. The representatives who sign the blocks have more than one percent of all the votes that they will take under the council. And the representatives who need to sign the upcoming blocks in the network are selected from the council (Yavuz et al. 2018).

The representative with any reason who misses their chance to sign the block in the network will need to leave the council due to the decrease in the votes that they need. Then, another representative who is willing to take the position of the previously left representative will be taken. The example of a DPOS algorithm is the EOS. The participation in the DPOS system is based on the permission.

The main advantages of a DPOS system are:

- It helps in saving the costs that are being spent on energy.
- It promotes decentralization.
- Most of the unwanted work is reduced for selecting the upcoming voter, which is a disadvantage in the POS systems.
- The owners have a chance to cast their votes without turning in over their actual resources.
- The DPOS systems are more scalable.
- The number of transactions per second is more when compared with those of POW and POS systems.

The disadvantage is that the required amount of decentralization will not be achieved. Figure 3.8 projects working of a DPOS system.

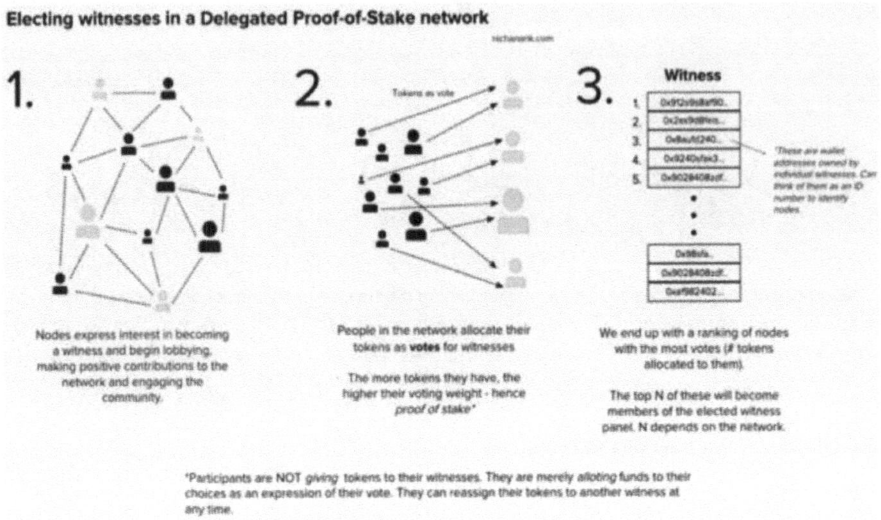

FIGURE 3.8 WORKING OF A DPOS SYSTEM.

Source: Coin Bureau

3.8.9 Noncompetitive Consensus

These algorithms process only one agreement at a time in the trusted network. All the nodes that are in the network check the agreements and reply to them. Yet, there is a disadvantage to these systems such as if any node is a malicious one and if it participates in the network more than a limited number of times, the system cannot be tolerated. Some of the examples of such types of systems are Paxos, Raft, and Practical byzantine fault-tolerant algorithms.

3.9 BENEFITS OF BLOCKCHAIN-BASED E-VOTING SYSTEM

1. To check if there has been any tampering in the data, the blockchains store the data securely using cryptographic algorithms. Each and every vote will be recorded or stored accurately, securely, and permanently as well. Therefore, no one will be able to modify the data.
2. The blockchain protects the user's anonymity or privacy, and it is also open to the inspection publicly. While there is nothing such as being completely secure, the tampering of data is impossible with the blockchain systems.
3. The E-voting system increases the level of the participation of the users. This is because the voting can be done by the voter from any place with any device which has an active Internet connection.
4. E-voting provides a chance for the voters where all the voters will be participating in the elections on the same terms. In the general election scenario, if a user is not able to travel from abroad, he wouldn't be able to cast his vote and the same goes for any user with disabilities. Such users' votes

will not be counted as they had not cast their vote. But, these problems can be eliminated by introducing E-voting where the users who are not able to travel can cast their vote from their willing location only, and the same is applicable for the users with disabilities.

5. The counting of the voting results will be way faster when compared to those with the traditional method of counting. And the trust will be increased in the election results because there cannot be a human error when it is counted through a system.
6. The marking of the votes and submission will be a lot easier as they help avoid errors like making the wrong selection, and this can be modified by the user before he/she submits their vote. There is no need for ballots in these systems.
7. The complete voting process will be auditable, and a receipt will be given to the user who had submitted their votes on their devices. Since blockchain is being used, the administrators can assure the users that their vote will be counted correctly.
8. The blockchain-based E-voting systems will reduce the fraud because the chances for the intentional variations during the vote count will be reduced.

3.9.1 Challenges for Blockchain-Based E-Voting System

1. Securing the users' data digitally: All the voters who are eligible to vote need to register before the election process starts. The information of the voters need to be available in the digital format, and the identity of the individual users need to be kept private.
2. Casting the votes anonymously: After the submission of a vote by the voter in the blockchain-based E-voting system, the users' vote need to be kept anonymous to all the members including the administrators of the system. So, the details of who might get the highest majority in the elections will not be available till the completion of the election process.
3. After submitting the vote, how the vote will be represented in the database is still unknown. But, we cannot store the data as simple as a plain text which will be a bad idea. So, we can use a hashed token to the data which provides both the integrity and anonymity.
4. The voters who had submitted their vote should be able to locate their corresponding vote to verify whether the system has taken their vote correctly or detect any malicious activity that happened to the votes of users during their voting process. This improves the thrust to the voters on the blockchain-based E-voting systems.
5. When we look at the costs, the cost for an E-voting system is lower when compared to those of the traditional voting systems, which require a huge amount of manpower and the cost as well. But, the setting up of the blockchain-based E-voting system is expensive in the initial phase.
6. By making the voting process online, the major challenge is the cyberattacks. During an election process these attacks are a major threat, and it is difficult to catch the culprit. Since most of the voters vote their votes

from either through their mobile phone or through laptop or computer systems which are easily hackable, any hacker can change the voters' vote and can remove any evidence that the corresponding users' vote has been compromised.
7. To overcome these risks, there are some techniques which can be used by preventing the deletion of the evidence that the attack has happened and transparency of the votes including the voter's privacy.
8. Since timing plays a key role in the election process, the infrastructure and the performance of the system should be of highest quality so that the remote voting will also be synchronous.

3.10 SECURITY ANALYSIS AND LEGAL ISSUES

While the present voting systems are not perfect, there are possibilities of security risks in the blockchain-based E-voting system, and with the increase in the usage of the Internet in almost all domains the possibility of the attacks also increases. Some of the possible attacks and the security-related issues need to be considered in the E-voting system so that the results of the votes do not get tampered within the middle.

3.10.1 Possible Attacks on Blockchain Network

Since currently, we are using blockchain as a main source for storing and accessing the required data which uses a decentralized way which is more secure. Although the blockchain technology can prevent many attacks, it is vulnerable to some attacks like the distributed denial of service attack, routing attacks, and Sybil attacks.

3.10.1.1 Distributed Denial of Service

The distributed denial of service (DDOS) attacks are difficult to carry out on a blockchain network, but the attacks are still possible. When the attacker attacks the blockchain network using DDOS, the attacker aims to slow down the server by making use of all the processing resources with multiple requests. The DDOS attackers intend to detach all the services that the blockchain provides. The DDOS can hack the blockchain using the DDOS botnets at the application layer.

3.10.1.2 Routing Attacks

The routing attack can affect both the whole network and the individual nodes. The main purpose of this type of attack is to disrupt the data transactions before sending them to the peers. It is almost nonviable for the other nodes to note this disruption, as the attacker partitions the network that is not able to communicate with each other. There are two different types of attacks in the routing attacks which are:

1. The partition attack, which partitions the network nodes into different groups
2. The delay attack, which disrupts the messages and sends them over the network

3.10.1.3 Sybil Attacks

Sybil attack is organized by giving several points to the identical node. The networks do not have valid nodes, and all the calls are transmitted to the different nodes. These attacks are mainly against the centralized networks. At the time of the attack, the attacker takes many nodes under his control that are in the network. Then, the victim is surrounded by many false nodes that end up with all their transmissions. At the end, the victim is exposed to double attacks. The Sybil attacks are very difficult to detect and prevent, but the following steps can work: increasing the cost of building new identities; requiring some form of trust by joining the network; or gaining user power based on the reputation (Zhang et al. 2020).

3.10.2 Anonymity

Anonymity means to keep a user identity unknown. In the blockchain-based E-voting system, only the user's public key will be made available which has also been hashed earlier. It makes sure that no one within the blockchain is able to identify any voters by removing their actual identity.

3.10.3 Confidentiality

Confidentiality means preventing sensitive data from reaching the unauthorized users. The data encryption is the most common method that is being used for maintaining the confidentiality of a user. The data that is being encrypted in the voting process is the voter's identity and the vote the voters' vote, and it will be made visible only when the election is completed.

3.10.4 Ballot Manipulation

In a normal paper ballot voting system, there is a possibility of a user voting more than once which affects the results of the elections. But in the case of ballot manipulation in the blockchain-based E-voting system, the user who tries to vote the second time, based on the approval of the peer nodes, will be rejected, and the vote will not be added to the ballot block.

3.10.5 Transparency

In today's voting system, there is no available method which can provide transparency to the voters. When a voter places his/her vote on the ballot, there is no assurance provided to the voter whether his/her vote has been calculated correctly or not. Any user vote can be changed owing to the errors made by the humans, such as the counter who counts the votes dislikes the party that the user had voted for, so there can be a misuse of the vote. There will be no valid data available that the vote had come from a particular ballot, so transparency in today's selection process is a must.

In blockchain, each and every transaction that is being made is made transparent for the verification since it is a distributed ledger. As all the transactions are being kept transparent, this makes sure that no fraudulent activities can take place secretly. Therefore, the accuracy is obtained through the blockchain because of its transparency.

3.10.6 Auditability

The results are calculated only after the election process is completed, and the complete process is made available as the blockchain keeps the record of each transaction. The activities that are fraudulent can be used to analyze how many transactions are fraudulent. The service that is used for the E-voting through blockchain is the smart contract in which the codes cannot be changed because the code is written on the blockchain permanently (Zhou et al. 2020).

3.10.7 Nonrepudiation

Nonrepudiation means that some person or user cannot deny something. That is, the outcome of the elections that is determined finally cannot be framed as unfair as each transaction is being kept transparent for the majority of the network.

3.11 CONCLUSION

There are different ways of voting that are being used for elections, but the blockchain technologies which are currently available are secure, and each transaction is transparent. However, there are some countries that conduct their elections through E-voting which is in a centralized environment which is not secure as the attackers can attack any centralized system keeping in mind that all the E-voting systems that are being developed based on the centralized system need to be stopped. To overcome these insecurities, we can use the blockchain technology and implement the E-voting system in a decentralized environment. The blockchain technologies work through openness and distribution which is why an E-voting system based on blockchain technologies is secure and the data cannot be altered. We can make use of the smart contracts which run in a localized environment. The implementation of smart contracts makes sure that the code is not altered by any one which helps in increasing the trust among the users regarding the E-voting. By using the blockchain-based E-voting, the users' privacy will be secured and the complete election process will be auditable whenever required, and no one can deny the result obtained through the blockchain-based E-voting since there is no possibility of tampering with the data.

But based on the current situations of the users and from the government's perspective as well, it seems that it won't be soon that the current voting systems will be replaced by the E-voting system. It will be possible only if the governments try to explore the blockchain technologies and invest in them to convert the current voting systems to E-voting which is mainly based on transparency. After accepting the blockchain technologies in the voting system, the government should educate the citizens about them to make use of these technologies in a better way.

REFERENCES

Abuidris, Yousif, Abdelrhman Hassan, Abdalla Hadabi, and Issameldeen Elfadul. "Risks and Opportunities of Blockchain based on E-Voting Systems." In *2019 16th International Computer Conference on Wavelet Active Media Technology and Information Processing*, pp. 365–368. IEEE, 2019.

Adeshina, Steve A., and Adegboyega Ojo. "Maintaining Voting Integrity using Blockchain." In *2019 15th International Conference on Electronics, Computer and Computation (ICECCO)*, pp. 1–5. IEEE, 2019.

Adiputra, Cosmas Krisna, Rikard Hjort, and Hiroyuki Sato. "A Proposal of Blockchain-based Electronic Voting System." In *2018 Second World Conference on Smart Trends in Systems, Security and Sustainability (WorldS4)*, pp. 22–27. IEEE, 2018.

Alam, Asraful, SM Zia Ur Rashid, Md Abdus Salam, and Ariful Islam. "Towards Blockchain-based E-Voting System." In *2018 International Conference on Innovations in Science, Engineering and Technology (ICISET)*, pp. 351–354. IEEE, 2018.

Al-madani, Ali Mansour, Ashok T. Gaikwad, Vivek Mahale, and Zeyad AT Ahmed. "Decentralized E-voting System based on Smart Contract by using Blockchain Technology." In *2020 International Conference on Smart Innovations in Design, Environment, Management, Planning and Computing (ICSIDEMPC)*, pp. 176–180. IEEE, 2020.

Anilkumar, Vysakh, Joseph Antony Joji, Asif Afzal, and Reshma Sheik. "Blockchain Simulation and Development Platforms: Survey, Issues and Challenges." In *2019 International Conference on Intelligent Computing and Control Systems (ICCS)*, pp. 935–939. IEEE, 2019.

Anjan, Spurthi, and Johnson P. Sequeira. "Blockchain Based E-Voting System for India Using UIDAI's Aadhaar." *Journal of Computer Science Engineering and Software Testing*, vol. 5, no. 3, pp. 26–32, 2019.

Ayed, Ahmed Ben. "A Conceptual Secure Blockchain-based Electronic Voting System." *International Journal of Network Security & Its Applications*, vol. 9, no. 3, pp. 1–9, 2017.

Bellini, Emanuele, Paolo Ceravolo, and Ernesto Damiani. "Blockchain-based E-Vote-as-a-Service." In *2019 IEEE 12th International Conference on Cloud Computing (CLOUD)*, pg. 484–486. IEEE, 2019.

Bera, B., A.K. Das, M. Obaidat, P. Vijayakumar, K.F. Hsiao, and Y. Park. "AI-Enabled Blockchain-Based Access Control for Malicious Attacks Detection and Mitigation in IoE." IEEE Consumer Electronics Magazine, 2020.

Bohli, J.M., J. Muller-Quade, and S. Rohrich. "Bingo Voting: Secure and Coercion-Free Voting Using a Trusted Random Number Generator." In *Proceedings of the 1st International Conference on E-voting and Identity*, pp. 111–124. Berlin, Heidelberg: Springer-Verlag, 2007.

Bulut, Rumeysa, Alperen Kantarcı, Safa Keskin, and Şerif Bahtiyar. "Blockchain-based electronic voting system for elections in turkey." In *2019 4th International Conference on Computer Science and Engineering (UBMK)*, pp. 183–188. IEEE, 2019.

Canessane, R. Aroul, N. Srinivasan, Abinash Beuria, Ashwini Singh, and B. Muthu Kumar. "Decentralised Applications Using Ethereum Blockchain." In *2019 Fifth International Conference on Science Technology Engineering and Mathematics (ICONSTEM)*, vol. 1, pp. 75–79. IEEE, 2019.

Casado-Vara, Roberto, and Juan M. CoRCHaDo. "Blockchain for Democratic Voting: How Blockchain Could Cast of Voter Fraud." *Oriental Journal of Computer Science and Technology*, vol. 11, no. 3, 2018.

Chafe, Shreya Shailendra, Divya Ashok Bangad, and Harsha Sonune. "Blockchain-Based E-Voting Protocol." In *ICT Systems and Sustainability*, pp. 245–255. Singapore: Springer, 2021.

Chaudhari, Ketulkumar Govindbhai. "E-voting System using Proof of Voting (PoV) Consensus Algorithm using BlockChain Technology." *International Journal of Advanced Research in Electrical, Electronics and Instrumentation Engineering*, vol. 7, no. 11, pp. 4051–4055, 2018.

Chaum, D., P.Y.A. Ryan, and P.Y.A. Schneider. "A Practical Voter-Verifiable Election Scheme." In *Proceedings of the 10th European Conference on Research in Computer Security*, pp. 118–139. Berlin, Heidelberg: Springer-Verlag, 2005.

Cruz, Jason Paul, and Yuichi Kaji. "E-voting System based on the Bitcoin Protocol and Blind Signatures." *IPSJ Transactions on Mathematical Modeling and Its Applications*, vol. 10, no. 1, pp. 14–22, 2017.

Dalia, K., R. Ben, Y.A. Peter, and H. Feng. "A Fair and Robust Voting System by Broadcast." *5th International Conference on E-voting*, 2012.

Garg, Kanika, Pavi Saraswat, Sachin Bisht, Sahil Kr Aggarwal, Sai Krishna Kothuri, and Sahil Gupta. "A comparative analysis on e-voting system using blockchain." In *2019 4th International Conference on Internet of Things: Smart Innovation and Usages (IoT-SIU)*, pp. 1–4. IEEE, 2019.

Han, Gang, Yannan Li, Yong Yu, Kim-Kwang Raymond Choo, and Nadra Guizani. "Blockchain-based Self-Tallying Voting System with Software Updates in Decentralized IoT." *IEEE Network*, vol. 34, no. 4, pp. 166–172, 2020.

Hanifatunnisa, Rifa, and Budi Rahardjo. "Blockchain based E-Voting Recording System Design." In *2017 11th International Conference on Telecommunication Systems Services and Applications (TSSA)*, pp. 1–6. IEEE, 2017.

Hao, F., P.Y.A. Ryan, and P. Zielinski. "Anonymous Voting by Two-round Public Discussion." *IET Information Security*, vol. 4, no. 2, pp. 62–67, June 2010.

Hjálmarsson, Friðrik Þ., Gunnlaugur K. Hreiðarsson, Mohammad Hamdaqa, and Gísli Hjálmtýsson. "Blockchain-based E-Voting System." In *2018 IEEE 11th International Conference on Cloud Computing (CLOUD)*, pp. 983–986. IEEE, 2018.

Jafar, Uzma, and Mohd Juzaiddin Ab Aziz. "A State of the Art Survey and Research Directions on Blockchain Based Electronic Voting System." In *International Conference on Advances in Cyber Security*, pp. 248–266. Singapore: Springer, 2020.

Karuppiah, M., A.K. Das, X. Li, S. Kumari, F. Wu, S.A. Chaudhry, and R. Niranchana. "Secure Remote User Mutual Authentication Scheme with Key Agreement for Cloud Environment." *Mobile Networks and Applications*, vol. 24, no. 3, pp. 1046–1062, 2019.

Karuppiah, M., and R. Saravanan. "A Secure Authentication Scheme with User Anonymity for Roaming Service in Global Mobility Networks." *Wireless Personal Communications*, vol. 84, no. 3, pp. 2055–2078, 2015.

Karuppiah, M., and R. Saravanan. "A Secure Remote User Mutual Authentication Scheme using Smart Cards." *Journal of Information Security and Applications*, vol. 19, no. 4–5, pp. 282–294, 2014.

Khan, Kashif Mehboob, Junaid Arshad, and Muhammad Mubashir Khan. "Secure Digital Voting System based on Blockchain Technology." *International Journal of Electronic Government Research (IJEGR)*, vol. 14, no. 1, pp. 53–62, 2018.

Kirillov, Denis, Vladimir Korkhov, Vadim Petrunin, Mikhail Makarov, Ildar M. Khamitov, and Victor Dostov. "Implementation of an E-Voting Scheme Using Hyperledger Fabric Permissioned Blockchain." In *International Conference on Computational Science and Its Applications*, pp. 509–521. Cham: Springer, 2019.

Kshetri, Nir, and Jeffrey Voas. "Blockchain-enabled E-Voting." *IEEE Software*, vol 35, no. 4, pp. 95–99, 2018.

Kumar, Sumit, N. Darshini, Sudhanshu Saxena, and P. Hemavathi. "VOTEETH: An E-voting System using Blockchain." *International Research Journal of Computer Science*, vol. 6, no. 6, pp. 11–18, 2019.

Liefhebber, W., and M. vd Laan. "Defining an Architecture for Blockchain E-voting Systems." Utrecht University, Bachelor thesis-Information Science 7 (2017).

Lin, C., D. He, N. Kumar, X. Huang, P. Vijayakumar, and K.K.R. Choo. 2019. "Homechain: A Blockchain-based Secure Mutual Authentication System for Smart Homes." *IEEE Internet of Things Journal*, vol. 7, no. 2, pp. 818–829, 2019.

Liu, J., X. Li, Q. Jiang, M.S. Obaidat, and P. Vijayakumar. "Bua: A Blockchain-based Unlinkable Authentication in Vanets." In *ICC 2020–2020 IEEE International Conference on Communications (ICC)* (pp. 1–6). IEEE, 2020 June.

Mitra, Mousumi, and Aviroop Chowdhury. "A Modernized Voting System Using Fuzzy Logic and Blockchain Technology." *International Journal of Modern Education and Computer Science*, vol. 12, no. 3, pp. 17, 2020.

Olaniyi, Olayemi M., Taliha A. Folorunso, Ahmed Aliyu, and Joseph Olugbenga. "Design of Secure Electronic Voting System using Fingerprint Biometrics and Crypto-watermarking Approach." *International Journal of Information Engineering and Electronic Business*, vol. 8, no. 5, p. 9, 2016.

Pandey, Archit, Mohit Bhasi, and K. Chandrasekaran. "VoteChain: A Blockchain Based E-Voting System." In *2019 Global Conference for Advancement in Technology (GCAT)*, pp. 1–4. IEEE, 2019.

Patidar, Kriti, and Swapnil Jain. "Decentralized E-Voting Portal Using Blockchain." In *2019 10th International Conference on Computing, Communication and Networking Technologies (ICCCNT)*, pp. 1–4. IEEE, 2019.

Patil, Harsha V., Kanchan G. Rathi, and Malati V. Tribhuwan. "A Study on Decentralized e-Voting System using Blockchain Technology." *International Research Journal of Engineering and Technology*, vol. 5, no. 11, pp. 48–53, 2018.

Pawade, Dipti, Avani Sakhapara, Aishwarya Badgujar, Divya Adepu, and Melvita Andrade. "Secure Online Voting System Using Biometric and Blockchain." In *Data Management, Analytics and Innovation*, pp. 93–110. Singapore: Springer, 2020.

Pawlak, Michał, and Aneta Poniszewska-Marańda. "Blockchain E-Voting System with the Use of Intelligent Agent Approach." In *Proceedings of the 17th International Conference on Advances in Mobile Computing & Multimedia*, pp. 145–154. 2019.

Poniszewska-Marańda, Aneta, Michaę Pawlak, and Jakub Guziur. "Auditable Blockchain Voting System: The Blockchain Technology toward the Electronic Voting Process." *International Journal of Web and Grid Services*, vol. 16, no. 1, pp. 1–21, 2020.

Punithavathi, P., S. Geetha, M. Karuppiah, S.H. Islam, M.M. Hassan, and K.K.R. Choo. "A Lightweight Machine Learning-based Authentication Framework for Smart IoT Devices." *Information Sciences*, 484, pp. 255–268, 2019.

Roh, Chang-Hyun, and Im-Yeong Lee. "A Study on Electronic Voting System using Private Blockchain." *Journal of Information Processing Systems*, vol. 16, no. 2, pp. 421–434, 2020.

Sadia, Kazi, Md Masuduzzaman, Rajib Kumar Paul, and Anik Islam. "Blockchain-based Secure E-Voting with the Assistance of Smart Contract." In *International Conference on Blockchain Technology 2019*, pp. 161–176. Singapore: Springer, 2020.

Shahandashti, S.F., and F. Hao. *DRE-ip:A Verifiable E-Voting Scheme Without Tallying Authorities*. Cham: Springer International Publishing, pp. 223–240, 2016.

Singh, Ashish, and Kakali Chatterjee. "Secevs: Secure Electronic Voting System using Blockchain Technology." In *2018 International Conference on Computing, Power and Communication Technologies (GUCON)*, pp. 863–867. IEEE, 2018.

Suralkar, Sunita, Sanjay Udasi, Sumit Gagnani, Mayur Tekwani, and Mohit Bhatia. "E-Voting Using Blockchain with Biometric Authentication." *International Journal of Research and Analytical Reviews*, vol. 6, no. 1, pp. 72–81, 2019.

Taş, Ruhi, and Ömer Özgür Tanrıöver. "A Systematic Review of Challenges and Opportunities of Blockchain for E-voting." *Symmetry*, vol. 12, no. 8, pp. 1328, 2020.

Vo-Cao-Thuy, Linh, Khoi Cao-Minh, Chuong Dang-Le-Bao, and Tuan A. Nguyen. "Votereum: An Ethereum-based E-Voting System." In *2019 IEEE-RIVF International Conference on Computing and Communication Technologies (RIVF)*, pp. 1–6. IEEE, 2019.

Wu, F., L. Xu, X. Li, S. Kumari, M. Karuppiah, and M.S. Obaidat. "A Lightweight and Provably Secure Key Agreement System for a Smart Grid with Elliptic Curve Cryptography." *IEEE Systems Journal*, vol. 13, no. 3, pp. 2830–2838, 2018.

Yavuz, Emre, Ali Kaan Koç, Umut Can Çabuk, and Gökhan Dalkılıç. "Towards Secure E-Voting using Ethereum Blockchain." In *2018 6th International Symposium on Digital Forensic and Security (ISDFS)*, pp. 1–7. IEEE, 2018.

Zhang, Jingyu, Siqi Zhong, Tian Wang, Han-Chieh Chao, and Jin Wang. "Blockchain-based Systems and Applications: A Survey." *Journal of Internet Technology*, vol. 21, no. 1, pp. 1–14, 2020.

Zhou, Yuanjian, Yining Liu, Chengshun Jiang, and Shulan Wang. "An Improved FOO Voting Scheme using Blockchain." *International Journal of Information Security*, vol. 19, no. 3, pp. 303–310, 2020.

4 The Efficacy of AI and Big Data in Combating COVID-19

Muralidhar Kurni, Saritha K, and Mujeeb Shaik Mohammed

CONTENTS

DOI: 10.1201/9781003152392-4

4.1 INTRODUCTION

The new coronavirus-19 (COVID-19/SARS-CoV-2) influenced healthcare and lives of people, education, transport, politics, and the supply chainsignificantly and changed the atmosphere completely (Q. V. Pham et al. 2020). The infected persons generally develop respiratory illnesses, and can recover with adequate and appropriate treatments. COVID-19, however, is a more dangerous entity, capable of spreading dangerously faster than other coronavirus families; this is how COVID-19 has become highly effective in human-to-human transmissions.

The World Health Organization and the Center for Disease Control and Prevention have already issued a set of public recommendations and scientific guidelines as war leaders on the new coronavirus (A. Abdulla et al. 2020). The COVID-19 risks are expected to be significantly reduced by cooperation between governments and major companies. Important information such as coronavirus maps, recent statistics, and most common COVID-19 queries is presented on the portal "www.google.com/covid19." Another platform that can help is the Supercomputer Architecture for coronavirus-related research developed by IBM, Amazon, Google, and Microsoft (A. Alimadadi et al. 2020). Moreover, many publishers also have free access to COVID-19 virus literature, technical standards, and other materials in light of the pandemic. In turn, arXiv, medRxiv, and bioRxiv web archive services quickly connect to all collected COVID-19 preprints (A. Banerjee et al. 2020).

Different applications, such as AI in computer science, AI in agriculture, AI in banking, and AI in health, include AI and big data. This technology is expected to play a leading role in combating the COVID-19 pandemic globally. The main purpose of this chapter is to show how efficient big data and AI are in the fight against COVID-19 and to explore cutting-edge technology. We also offer a case study about India using AI and big data to battle COVID-19.

4.2 COVID-19 PANDEMIC

The source of COVID-19 is a severe acute respiratory syndrome, coronavirus 2 (SARS-CoV-2) (A. López-Cortés et al. 2020). On December 31, 2019, in Wuhan, China's Province of Hubei, the first infected case of COVID-19 was reported, and it later spread to most countries around the world. The number of new cases is also very high. Due to the global pandemic, the World Health Organization's risk assessment for COVID-19 has risen to the greatest extent.

Since COVID-19 has changed the world profoundly, attempts have been made to provide solutions to the COVID-19 epidemic. Government measures are mainly responsible for halting the pandemic, such as closing the (partial) area to limit infection transmission, allowing the health system to handle and execute crisis-related mitigation packages, and enforcing COVID-19 adaptive policies. At the same time, it is advised to remain secure and protect others by giving advice such as using a mask on public sites, washing of hands, maintaining social distance policies, and the providing the provision of latest symptom information to regional health institutes.

Conversely, various stakeholders such as government, business, and academics have now accorded priority and particular attention to COVID-19-related research and development. Researchers have demonstrated essential implications for the COVID-19 global pandemic supply chain, including feasibility, stability, robustness, and resilience as a range of supply chain aspects. Computer scientists have worked for COVID-19 and a worldwide campaign to develop effective coronavirus vaccines and medical treatment. Motivated by the huge success of AI and big data in a variety of areas, we present the state-of-the-art COVID-19 coronavirus disease approaches and solutions.

4.3 COVID-19 AND AI

Over the years, psychologists, roboticists, and other experts have predicted AI powers and what can be done with its assistance. The scientific community is attempting to find trends in clinical trials and studies of the COVID-19 outbreak worldwide so that the existence of the virus can be anticipated well in advance. The instruments that researchers explore for this mission are machine learning (ML), deep learning, and AI, resulting in a certain degree of success.

Cranfield University students in the United Kingdom developed computer models to classify COVID-19 in radiographic images (A. N. Islam et al. 2020). The models are used to interpret chest X-ray images by computer vision and AI. As a result, you can categorize details that would usually not be identified by your naked eye and help diagnose COVID-19.

CORD-19, an open research dataset of COVID-19 is free for researchers to work on new developments in AI and NLP (natural language processing). The data are also available to AI practitioners to work out and develop technologies that can directly assist and alleviate the burden on our healthcare professionals.

AWS (Amazon Web Services) has created a platform for ML (https://cord19.aws/), which allows a researcher to search for COVID-19 research papers and obtain answers. Ramco Systems, a Singapore-based software and services provider, has prepared itself for this stage of COVID-19 by introducing face recognition and thermal scanning in their working room (A. Prakash et al. 2020). The first step in this new system is recognizing the face, which is the next step after authentication by thermal scans, in which the employee can enter only when the temperature is normal. Gurugram-based AI start-up Staqu Technologies has developed technology to recognize persons who do not meet social distance standards. A French start-up Clevy.io launched public chatbots to scan for COVID-19-related government communications. At the last, Chan ZuckerbergBiohub developed a model that could estimate the number of undetected cases of COVID-19. This model used ML to quantify the number of undetected cases by evaluating the virus mutation transmitted in the population.

Benevolent AI uses AI Drug Detection Tool to track the efficiency of currently certified drugs in eliminating COVID-19 spread. Moreover, in the course of COVID-19 scientific studies, the available literature for studying the virus is expanding, and researchers have more data available. BlueDot, a starter in Canada in the area of AI, is among those technology companies that had warned of the outbreak of flu-like disease even in the run-up to the WHO study. BlueDot used an AI-based algorithm to detect the spread of the pandemic by going through different international media reports and air-ticketing patterns (A. Riccardi et al. 2020). Jointly, the University of Johns Hopkins and ETH Zurich have developed an interactive dashboard to track COVID-19 infections across the globe. The AI company Vehant Technology developed by the IIT Delhi company FebriEye, a thermal screening and a camera that tracks social distancing, facial masks, is a similar technique to Ramco System. In addition, India has established a government application called Aarogya Setu to alert people when/if they are around a person infected with COVID-19.

4.4 HOW TO FIGHT CORONAVIRUS WITH THE HELP OF AI?

By March 3, 2020, COVID-19 had killed 3,168 people, with 92,880 plus confirmed cases of infected persons in at least 79 countries. Therefore, the overall mortality rate of the coronavirus is higher than the death rate of its "cousin" SARS virus in 2003 and the "bird flu" virus in 2013. Experts caution us that wearing face masks will not avoid spreading COVID-19. But advances in AI and Genetic Applied Science made it simpler, quicker, and more affordable to understand the spread and management of the virus and the destructive consequences (A. S. S. Rao et al. 2020).

1. *AI may predict epidemics*: AI will alert us to a future epidemic and give us sufficient time to prepare for it. BlueDot, a global AI database company, uses an AI-driven algorithm to analyze information from multiple sources and monitor hundreds of infectious diseases. In addition, BlueDot uses natural-language processing and ML. On December 31, 2019, Blue Dot alerted the World Health Organization (WHO) and the US Centers for Disease Control and Prevention (CDC) to avoid visiting Wuhan. Moreover, BlueDot forecasted earlier in March 2020 that other Asian city outbreaks could occur by analyzing routes of travel and flight. In the future, AI might also predict human actions and possible outbreaks by using social media data.
2. *AI will speed up the drug development and discovery*: AI is not only able to alert us of an impending epidemic but also can allow us more rapidly than ever to find, produce, and scale new drugs and vaccines. A quick and effective recreation of the virus genome sequence and a copy of the virus is the secret to creating a vaccine. In just 1 month, for example, Chinese scientists recreated the virus's genome sequence. In 2003, scientists had to reconstruct the genome sequence of the SAR virus to compare how remarkable it was. Researchers in Australia have developed a lab-grown of the virus from an infected patient. We can rapidly establish and confirm diagnostic tests with a precise genome sequence and a virus copy replica. Therefore, within three months from the rise of the pandemic, we have one vaccine candidate who

entered the clinical trial. Insilico Medicine, a Longevity Vision Fund portfolio company, used its AI system to classify thousands of potential drug molecules successfully in just four days. Insilico Medicine released updated results on its website so that all investigators could download their information for free, eventually adding to the overall efforts to fight the pandemic.

3. *AI may help manage disease outbreaks, reducing the burden of death*: AI can help reduce the death rate by reducing the burden on healthcare professionals by informing patients about proper care procedures. The risk of exposure and contraction to COVID-19 is high for healthcare professionals like doctors, nurses, and hospital personnel. This risk can be relieved by AI. For instance, China uses robots for faster diagnostic inspections, and AI helps city ambulances in Hangzhou speed up transport. In case we contract the virus, AI may also help remind us of what we should do. For example, China has introduced an app to help people monitor whether a confirmed coronavirus patient has taken a flight or train. At the same time, China uses drones to ensure that people take the right precautions. Thus, while the rapid global spread of COVID-19 is frightening, AI gives us a ray of hope for our healthcare leaders, policymakers, and public servants.

4.5 AI MAKING THE COVID-19 DRUG DEVELOPMENT CHEAPER, QUICKER, AND MORE EFFECTIVE

New drugs are slow and expensive to test. AI interrupts clinical trials from data collection, patient recruitment, and follow-up, and COVID-19 has catalyzed its adoption (A. SesagiriRaamkumar et al. 2020). In 2020, almost 5,000 clinical trials to test life-saving therapies and new coronavirus vaccines were initiated. Registrations for COVID-19 are 80% higher than the standard registration for clinical trials. However, this does not matter too much because fewer than 10% of qualified patients are involved in studying diseases such as cancer. Moreover, patients only take part in a clinical trial if no prior therapy has existed.

In addition, the eligibility cannot be determined by all diagnosed patients alone. Registered individuals are often involved in a trial as a time-consuming, expensive effort. The treatment is too ineffective for another player, as the total cost of drug testing for almost a decade is over $1B. For the $52B markets in clinical trials, a makeover is expected. Big techs and entrepreneurs actively build solutions for clinical trials, from IoT to remote monitoring, to HER's (electronic health records) ML to AI-based data protection systems. This segment identifies a patient's trajectory in a conventional clinical trial stage and addresses applications for new technologies such as AI in all phases. We mainly concentrate on drug-based trials although the technologies apply to different clinical trials.

4.5.1 Why Are Faster Trials Essential for Pharmaceutical Companies?

Marketing a drug is a lengthy and challenging task. Studies report that the clinical trial phase—in which experimental medicines are tested until approved by the

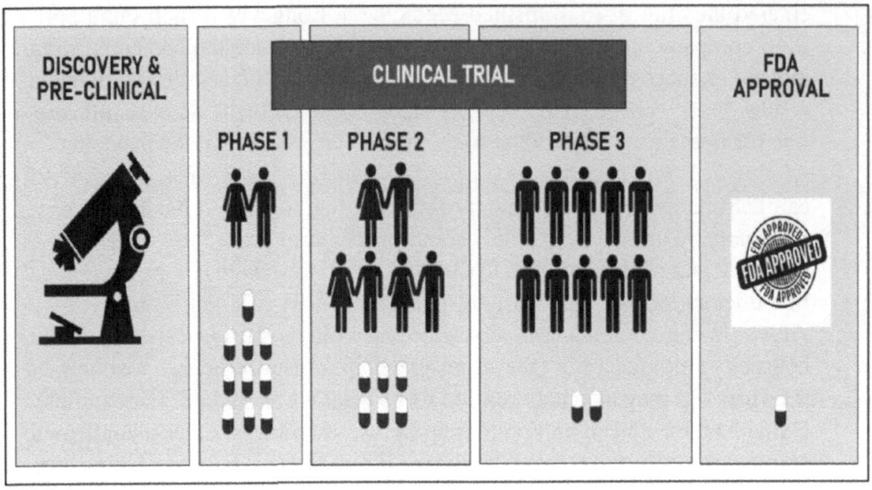

FIGURE 4.1 Phases involved in bringing a drug to market.

Source: Z. Li et al. (2020)

FDA in patients—takes nine years and averages $1.3B. The various phases involved in bringing a drug to market are represented in Figure 4.1. The multiple phases of clinical trials increase cost and uncertainty between Phase I and Phase III. While much time and money are spent on trials, only 1 out of 10 drugs enter the phase when the FDA will approve a clinical trial.

For different reasons, clinical trials fail, including failure to attract sufficient participants, mid-trial patients' dropout, side effects, and inaccurate results. In addition, of course, later-stage testing is more expensive. For example, from Switzerland, Novartis has attributed a 15-percent decrease in Q117's net income to a failed Phase III medication intended to treat heart failure. For biopharmaceutical start-ups, the cost of failure is higher. With minimal cash, a leading candidate for a start-up fails to succeed in a clinical trial. This is because IPO firms are seldom in late-stage clinical studies until at least one promising drug is discovered.

Recently, a development in SPAC has enabled businesses to increase the likelihood of surviving by earlier, higher-risk access to public capital. The high costs of clinical trials also impact patient costs downstream since the R&D costs of failed trials are linked with the prices of approved drugs to keep profitable by biopharma firms.

4.5.1.1 The State of the Clinical Trials

After the commercially available medication has not succeeded, registering, and participating in clinical trials, patients must follow a complex procedure. A typical patient journey is represented in Figure 4.2.

Many clinical trials use rudimentary data collection and control procedures, which also burden the patient—for example, sending medical records by fax, manual counting of remaining pills in bottles, and diary entries to assess adherence to medicinal products. This is a disruptive operation.

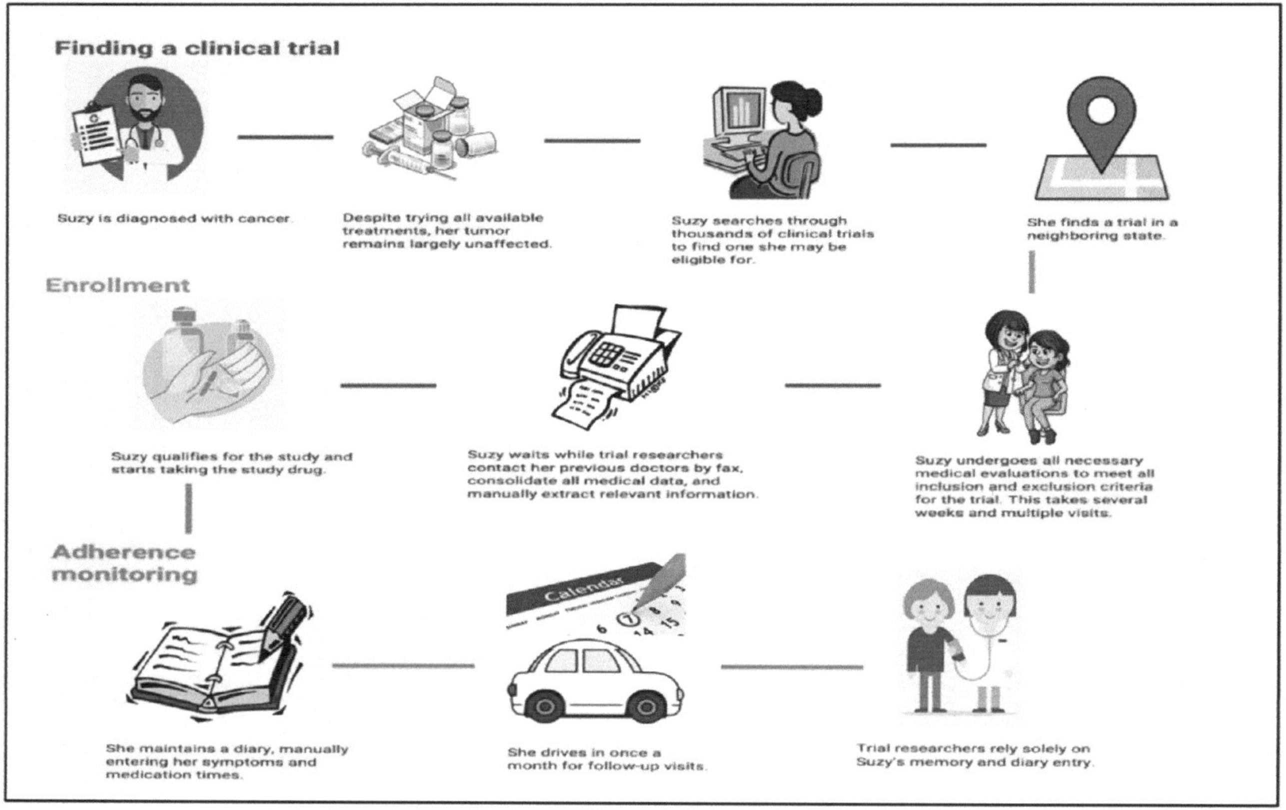

FIGURE 4.2 Understanding a patient's journey from diagnosis to enrolment to monitoring.

Source: Z. Li et al. (2020)

4.5.2 How AI Can Alter All Phases of Clinical Trials

AI will alter each stage of the clinical trials from testing to enrolment and adherence to drugs.

4.5.2.1 Clinical Trial Finding

The correct patient's correct test is for the clinical study team and the patient a time-consuming and demanding task. "Only 3% of cancer patients are currently registered in clinical trials.—WhiteHouse.gov, May 2018." Around 80% of clinical trials do not comply with registration deadlines, and about one-third of Phase III clinical trials are closed due to admission difficulties. There are currently more than 22,000 clinical trials recruiting patients in the United States. If a doctor knows of an ongoing trial, patients will sometimes receive trial recommendations from their physicians. Otherwise, it will also be the patient's responsibility to scour through ClinicalTrials.gov—an extensive government database of previous and current clinical trials.

NLP can assist with the extraction and analysis of relevant details from the EHR records of a patient, compare the eligibility requirements for ongoing trials, and suggest matching studies. In reality, it is one of the most sought-after AI applications in medical services to collect information from medical records—like EHRs and laboratory pictures. However, there are several barriers to accessing patient records, including unstructured healthcare information and disparate nonconnecting data sources.

4.5.2.1.1 The EHR Interoperability Challenge

No uniform format or central repository of patient-medical data is available despite the $27B federated incentive program for hospitals and providers to implement EHRs. In reality, access to the records of the hospital from all health facilities visited by patients is still difficult. Data sharing is permitted with patient consent under the Health Insurance Portability and Accountability Act (HIPAA). This provides AI start-ups with the possibility of evaluating medical data and proposing qualifying patients in minutes—a procedure that would take months otherwise. However, problems with the safe sharing and/or interoperability of health information between organizations and software systems remain. This problem was highlighted by the COVID-19 pandemic and attracted investor attention to the EHR ecosystem. EHRs' news reports were sky-rocketed too. The same EHR program cannot be used to enter data from various hospitals and providers that treat the same patient. Researchers are still requesting patient records by fax in several clinical trials, which usually return the data in the form of PDFs or photographs, including pictures of handwritten notes.

For AI, this is challenging. As one study from Harvard, MIT, Johns Hopkins, and NYU researchers point out, the standard natural language-processing tasks such as word sense disambiguation and sentiment analysis are difficult clinical notes that are misspelled. Flatiron Health, a Health AI corporation, describes this in a patent application: "Structured data can also become unstructured due to transmission methods. For example, a spreadsheet that is faxed or turned into a read-only document (such as PDF) loses much of its structure." This dated manual method makes collecting reliable data to assess a patient's eligibility difficult for clinical test investigators. Startups tackle from different perspectives the issue of patient recruitment (A. Wen et al. 2020).

Deep 6 AI uses NLP to collect clinical data from health records—including symptoms, diagnostics, and therapies. Its software can also classify patients with conditions not explicitly mentioned in EHR data and enhance the patient-clinical-test match rate. In its last fundraise, Deep 6 AI was estimated at more than $140M. Another solution is the marketplaces of clinical trials as offered by SubjectWell. The network of SubjectWell enables researchers to reach prescreened patients. A smaller community attempts to collaborate with a direct-to-consumer approach to interoperability hurdles. Clara Health, for instance, provides a patient-friendly approach for finding new options. The approach fits patients in studies and provides continuous assistance throughout the process and improves retention. The Founders Fund and Khosla Ventures back the enterprise. Established players from other healthcare sectors also join the recruiting field for clinical trials. 23andMe also offers recommendations on which studies are good for its 12M+ customers based on their genetics (A. Zhavoronkov et al. 2020).

Additional emerging consumer-focused recruitment solutions include:

- Social networking sites that are unique to a disease such as Be the Partner.
- Open-source applications such as Google Health Studies are available to researchers to create their software for study.

4.5.2.1.2 Acquisitions as a Method for Obtaining Patient Information

Flatiron Health solved the interoperability issue by acquiring Altos Solutions, a company focusing on oncology and electronic medical records (EMR). In that era, Flatiron sold its cloud analysis platform to healthcare and life sciences firms, and oncology companies such as Florida Cancer Specialists used altos' EMR. As a result, instead of depending exclusively on access to third-party EMRs, Flatiron directly accessed raw patient information.

In 2018, Roche bought Flatiron at $2B+ to access its real-world proof—observations from EHRs, claims, and wearable sensors. These data will be used to enhance Roche's construction of cancer pipelines. In the meantime, biopharmaceutical firms aim to partner for patient access. For instance, in November 2020, Janssen Pharmaceuticals partnered with Tempus for access and improvement of patient identification and clinical and molecular database.

4.5.2.2 Enrolment Challenges

Regrettably, enrolment problems are not over if a patient chooses a clinical trial. A preliminary phone screen must be completed and then the patient examined in person or remotely at any participating location to validate eligibility. Every trial also contains conditions for the inclusion and exclusion to be met by each patient. Unfortunately, these words are sometimes juxtaposed with medical jargon, as seen in a Phase II breast cancer study screenshot below. ClinicalTrials.gov has shown that the duration and cost of these trials started in November 2017 and are due to end in May 2025.

In this case, patients must be evaluated to satisfy both inclusion and exclusion requirements, such as laboratory and imaging tests. Depending on their availability and how far they reside from the test site, some patients finish these procedures in less than a week. However, the process could take many visits for other people with children, inflexible occupations, or long journeys.

Telehealth systems allow this process to be streamlined. For instance, in 2020, Deaconess Health System partnered with TytoCare to incorporate the associated examination tools of the start-up with Deaconess Clinic LIVE, the proprietary virtual care platform for the health system. The software and mobile app Tyto allowed quarantined patients to perform independent testing by collecting cardiovascular, lung, neuronal, ear, skin, and abdomen data. You can then share your results with remote doctors in real time.

The eligibility of platforms like TytoCare and the conduct of virtual clinical studies can prove essential. The patient, if willing, signs a consent document that agrees with the conditions of the test. This involves considering possible secondary consequences, readiness to provide biological samples, and any expenditure not covered by the research budget. AI solutions for extracting patient medical record information simplify the registration process by automating certain inclusion and exclusion requirements (A. Abdollahi et al. 2020).

4.5.2.3 Adherence to Drugs

After an analysis is completed, patients receive an experimental medication (or placebo). In the first course of treatment, patients go home with essential medications (e.g., a 30-day pill bottle with dose instructions) and a daily diary. Instead of electronic systems, several clinical trials continue to use paper diaries. Patients are asked to remember while taking the study medicine if other drugs had any adverse effects during those days (including headache, stomach ache, or muscle aches).

Inefficiencies are affected by this process:

1. *Patient memory reliance*: When a patient returns for check-in, the investigator examines the pill bottle to check for any blanks and anomalies in the patient's diaries. When the diary entries lack detail, the researcher relies on the patient's event memory.
2. *Obsolete recording system*: Paper records that could be missing information or lost are outdated and unsatisfactory ways of recording the most important test data points.
3. *Risk of withdrawal*: Frequent travel to a routine check-in clinical trial site has a negative impact on patients' time and resources, particularly for outpatients. This increases the likelihood of leaving.
4. *Additional payments*: Since the patient signing document includes out-of-pocket charges, many patients do not grasp the extent of the fees. For example, the study does not include additional MRI and laboratory testing during follow-up visits, and insurance insurers may not reimburse those tests since they are for research purposes and are not necessarily necessary for medical purposes.

Noncompliance may have adverse health effects, incur costs if new participants are to be recruited in a study, and the accuracy of the trial results interferes. In general, therapeutic effectiveness requires adherence rates of 80% or higher. But up to 50% of prescribed prescriptions are unfortunately mistaken. In response, the sponsors of clinical studies invest in new technologies to mitigate noncompliance (M. Anshari et al. 2019).

4.5.2.3.1 Visual, Auditory, and Digital Phenotyping

Some start-ups have visual drug management evidence. For example, patients at risk of noncompliance are identified by platforms such as AiCure using an Interactive Medical Assistant based on data collection. Patients use their phones to film a drug, and AiCure ensures that the right person has taken the right pill. In late 2016, Baird Capital raised a $24.5 million series C and partnered with Science 37 for virtual clinical trials in April 2020.

Digital phenotyping and speech analysis for drug adherence assessment are also new technologies. Mindstrong uses digital phenotyping to assess moods based on the way users communicate with their mobile devices. Mindstrong uses digital phenotyping technology. Subtle improvements in the patient's brain health are identified using Aural Analytics, which recently worked with MAH in the amyotrophic lateral sclerosis study.

4.5.2.3.2 AI and IoT for Remote Patient Monitoring

Some start-ups design their own monitoring equipment and sensors and then add a ML layer to interpret the data to allow remote patient monitoring for clinical trials. Others only build the AI software and integrate it with on-home monitoring systems from third parties.

Connected devices can allow drug adherence in real time. Optimize.health (formerly Pillsy) has, for example, introduced an intelligent drug bottle with a relevant cellular application that provides remainders, educational content, dose monitoring, and patient-reported data capability for providers. In August 2020, the firm raised a $15m series of a remote patient surveillance tool. Additional solutions concentrate on physiological data collection. For example, AliveCor's EKG applies real-time ML to detect irregular heart rhythmites, such as atrial fibrillation. In addition, the company cooperated with Medable in April 2020 to enable clinical studies focusing on decentralized cardiology. In November, AliveCor recently collected a $65M Series E round (B. Pirouz et al. 2020).

In the meantime, Sequoia has supported capital Biofourmis uses wearable equipment for tracking the vitalities and fitness of users on the AI app. The start-up collaborated with the University of Hong Kong, for instance, to identify subtle health changes and help accelerate virus quest in the temperature, oxygen levels, and heart rate in patients infected with COVID-19. In addition, payers, insurers, and pharmaceutical firms use Biofourmis' platform to educate decisions about treatment effectiveness and provide personal attention. In April 2020, the company acquired Gaido Health from Takeda Pharmaceutical to expand into oncology remote-monitoring services. On-site checks are offered in the form of wearables and AI, which allows continuous, real-time monitoring of physiological and behavioral changes in the patient.

4.5.3 How Big Tech Interrupts Clinical Trials

Not only start-ups but big technology companies also work on clinical trials. Big technology companies use their mobile devices to build clinical trial platforms. For example, Apple has built an ecosystem of clinical trials around the iPhone and Apple Watch since 2015, which allow for the collection of health information in real time. In addition, their open-source applications—ResearchKit and the CareKit—support the recruitment and remote monitoring of patients through clinical trials.

However, recently, Google was more spatially involved. Google builds an ecosystem of clinical research by using Google's Android Health Studies app and creating products through its Verily Life Sciences subsidiary. Google can provide medical researchers with sources of patient health data via these products.

4.5.3.1 Google's Healthcare Data Platform

Verily actually introduced the project Baseline in 2017 to encourage health studies through mapping. As a result, by mid-2019, Novartis, Sanofi, Pfizer, and Otsuka had worked in partnership with Verily to use their resources to make clinical trials more effective. The program was also partnered with the School of Medicine at Duke University, Stanford Medicine, and the American Heart Association.

To date, some achievements include an FDA-licensed EKG watch and an abnormal pulse monitor approved by the FDA. In addition, Google released a new Android application—Google Health Studies—in December 2020, which simplifies user involvement and clarifies how its data are used for health research.

The Boston Children's Hospital and the Harvard Medical School have also partnered in a 100,000-person acute respiratory disease trial with Google to enroll Android users. The analysis will use survey responses and mobility data to examine the transmission dynamics of pathogens such as COVID-19 (B. QIN et al. 2020).

4.5.3.1.1 Disrupting EHR Data Sharing

The Interoperability Readiness Program was introduced recently by Google to allow medical companies to understand the current data status and improve system-wide standardization and integration. In addition, Google launched its Cloud Healthcare API in April 2020 for healthcare systems and quickly signed into top medical centers like the Mayo Clinic. These measures align with Google's commitments to interoperability and data-sharing requirements in healthcare 2018 (also signed by IBM, Amazon, Salesforce, and Microsoft).

Google also works to bring its systems and data to the cloud with EHR suppliers, including Meditech. A possible result of these collaborations may be a two-way data flow in which EHR suppliers are encouraged to integrate patient-generated data with Google apps (B.R. Beck et al. 2020).

4.5.3.1.2 What Does This Data Mean for Clinical Trials?

Google has become the central element of the health data system with the widespread use of mobile devices, providing previously inaccessible real-time data and collecting information that is difficult to consolidate HER information. As a result, the opportunities for early diagnosis with AI and ML, decision-making for medication design, the right pool of patients for studies appear infinite, and patients track progress remotely.

Many studies have experimental group (patients taking the trial drug) and a control group (patients who get a placebo drug). A control group aims to provide a basis for comparing the symptoms of the experimental group. As with Project Baseline data, patient data will help build digital twins and reduce the need to establish a control group, thus reducing the recruitment bottlenecks.

4.5.3.2 The Moves of Apple and Facebook

For Apple, clinical trials have less of a strategic significance. Over 500 doctors and medical researchers have used the tools for studies involving more than 3M people in 3 years following the launch of their open-source software ResearchKit and CareKit. Apple remained a partner for clinical trials with pharmaceutical companies such as Johnson&Johnson.

Facebook, which released a Preventive Health tool at the end of 2019, is another technological company that can enter this space. Due to the depth of Facebook's personal data and the community's self-organization.

4.5.4 How COVID-19 Influenced Tech Adoption in Clinical Trials

The COVID-19 pandemic has catalyzed the introduction of innovations that can increase the performance and cost of clinical trials.

4.5.4.1 Study Design

The central theme as scientists struggle with COVID-19 has been adaptive design—which requires a more versatile approach to a trial. While conventional studies can be stiff on primary endpoints and dosing schemes before the next step is started, an adaptive design enables researchers to change these measures as trials progress.

The test by both Regeneron Pharmaceuticals and Sanofi for COVID-19 followed an adaptive design for antibody therapy. A randomized nondual-blind model also followed this model in the SOLIDARITY trial of the WHO. The use of open research forums to speed up future results and conclusions is another tendency of the COVID-19 study initiatives.

In partnership with DCM Ventures, Mendel.ai has developed a COVID-19 search engine that allows researchers to gather data for their respective studies. The COVID-19 Open AI Consortium was established by Owkin, which promotes collaboration in key research fields, including cardiovascular complications. Since clinical trials are microscopically designed during this global pandemic, this project may provide researchers learning opportunities since alternative approaches are challenged by conventional studies.

4.5.4.2 Virtual Trials

COVID-19 has sparked interest in telemedicine and remote-monitoring applications in decentralized or virtual clinical trials. The remote model does not comply with all clinical trials—for example, frequent diagnostic imaging or other individual evaluation procedures. This model offers, therefore, a chance to better handle patient involvement. The potential advantages of virtual trials include lower costs, a broader eligible patient network, and improved patient retention rates.

Science 37, which provides complete clinical studies services using its virtual Metasite model, is one company in this field. It builds on a network of researchers, mobile nurses, and study coordinators to make studies more available to patients. It has raised a total fund of almost $107 million from investors like Amgen Ventures, GV, the Novartis Venture Fund, and the Sanofi-Genzyme BioVentures.

A collaboration with Innovo Research was announced in April 2020 to reduce the amount of time taken to start a COVID-19 clinical trial. This initiative is planned by building on the national site and patient network of Innovo and implementing the platform of Science 37. New debates about how technology can solve major gaps in the clinical trial environment today have arisen with the new coronavirus. As dependency on these technologies increases, these software platforms could influence future studies.

4.6 AI FOR COVID-19 PANDEMIC: A SURVEY ON THE STATE OF THE ARTS

In this section, we review the state of the arts of AI for the COVID-19 pandemic.

4.6.1 Understanding the Virus

Knowing the virus and its proper study using AI methods in this field is an important challenge in managing the pandemic. The virus protein sequence was detected using the linear regression, KNN, and SVM by B. Wang et al. (2020). An intrinsic SARS-CoV-2 genomic signature is described by a ML alignment-free algorithm in C. Li et al. (2020). The mutation rate of SARS-CoV2 is investigated using the LSTM algorithm by C. S. Dule et al. (2020). A Siamese Neural Network is proposed by C. T. Mowery et al. (2020) to differentiate between SARS-CoV-2 viruses and HIV-1 and Ebola. An AI-based method is given in this section to determine the origin of the virus.

4.6.2 Monitoring the Pandemic

Some studies have used AI to track the pandemic and its consequences. D. Ivanov et al. (2020) estimated the pandemic effect on fatality, the number of individuals infected by the disease, and recovery in a hybrid cell automaton. By D. Ivanov et al. (2020), a ML algorithm was proposed to study the impact of temperature, humidity, and wind speed on the number of infected persons. The effect of the pandemic on tourism is discussed by D. M. Gysi et al. (2020). In this work, the pandemic properties are determined by a long short-term memory neural network.

4.6.3 Controlling the Pandemic

It is essential that the reproductive rate is small to manage the pandemic. This section summarizes studies using AI methods for controlling the pandemic and reducing the rate of infection.

- *Contact tracing*: ML techniques are used by D. Sonntag et al. (2020) to create a contact tracking program. In D. Tátrai et al. (2020), a ML simulation algorithm to track, prevent, and predict the spread of pandemics and the subsequent outbreak is also suggested.

- *Identifying COVID 19 cases*: A call-based dialog agent for active monitoring in Korea and Japan is developed (D. Yang et al. 2020). In D.Wang et al. (2020), the AI methodology for mobile assessment agents for epidemics is introduced as an integrated technique for mobilization. A low-cost self-test and tracking device coupled with AI for COVID-19 is proposed. The classification of COVID-19 cases implemented using mobile online surveys is provided with the ML algorithm. Symptoms like diarrhea, nausea, conjunctivitis, and loss of taste are employed to classify patients into distinct groups and discover their possible illnesses (E. Ong et al. 2020).
- *Infection testing*: AI methods are used in G. Chandra et al. (2020) to research the association between swab tests and reported infections with attention to the degree of sickness. In G. Giordano et al. (2020), it is proposed that mobile device social relationships may be used to monitor disease spread. In G. Goh et al. (2020), COVID-19 infection is tested for humans by analyzing immunochromatographic lateral flow assays (LFA) to provide an evidence structure that helps inform the pairing of the LFAs to obtain better classification via ML.
- *Risk evaluation*: An ANN in G.S. Randhawa et al. (2020) carries out the COVID-19 risk assessment in urban districts. The importance of using AI-based search tools and the argument that future research on the disease requires smart searching techniques is addressed in G. Shtar et al. (2020). The AI-based framework, which offers a hierarchical risk-based assessment at the group level to support the development of pandemic control strategies, is proposed using data from heterogeneous sources and AI algorithms (A. Haleem et al. 2020). A risk assessment in a given area is automatically predicted hierarchically by the system from the state, region, city, and place.
- *Social distancing*: A CNN is proposed in Haleem et al. (2020) to track people to obey the guidelines in public places. I. A. T. Hashem et al. (2020) proposes an algorithm of machine vision that uses AI methods to track people who do not follow laws of social distancing. A surveillance video can track social distancing through a deep learning framework. Social distancing policies in various countries argue that they have varying implications. To investigate this, a hybrid ML platform called SIRNET (J. Kim et al. 2020) is proposed. AI algorithms and machine vision are used in J. Stebbing et al. (2020) to track employees and identify violations. The identification of the facemask wearing conditions is proposed. Deep text classification models for classifying information from social media during the pandemic are provided (J. Zhu et al. 2020). The data were gathered using Facebook commentary analysis.
- *Understanding the pandemic*: In K. Arora et al. (2020), it was suggested that a deeper grasp of the actions of the COVID-19 pandemic could lead to the understanding of other outbreaks such as influenza. The authors investigate the application of DNN to conduct sentinel syndromes for COVID-19. The method is built upon auto-encoders 7 for aberration detection, which leverages the distribution of symptoms to differentiate between diseases. The argument in K. Avchaciov et al. (2020) is that AI systems could predict China's pandemic before surprisingly capturing the planet. The authors

investigate how early viral detection can be minimized using AI systems by examining viral outbreaks over the past 20 years.

- *Combating misinformation*: Social media is an effective forum for exchanging news and personal experiences and views in real time and internationally during the pandemic, helping people to build up their awareness of the condition and how they can face its problems. The nature of disinformation and social exhaustion, however, impair its utility. In K. Heiser et al. (2020), the simulation of structural equations and neural network techniques investigates how motivational and personal variables influence exhaustion in social networks. An online learning algorithm for the analysis of COVID-19 material connected with vaccination is used (K. S. Pokkuluri et al. 2020).
- *Policy suggestion*: A ML algorithm in K. Siau et al. (2020) is proposed to detect systemic breakdowns within positive case dynamics with territorial panel data to provide policy recommendations to tackle the disease. In [46], the behavior of the pandemic in governmental actions is modeled by an algorithm using neural networks. A decision optimization algorithm is then suggested.

4.6.4 Managing the Effects of the Pandemic

The pandemic has affected many facets of people's lives, economy, business, and so on. AI methods are employed in some studies to develop ways to handle the pandemic's impacts. We cover these studies in this section. The way AI approaches should handle disease-induced problems is overviewed in L. Bonacini et al. (2020 and L. Erlina et al. (2020).

- *Utilities*: The pandemic has caused unparallel problems to utility and grid operators. The lockdowns and constraints have changed the magnitude and trend of power consumption profiles worldwide. This has made load forecasting difficult. Traditional algorithms use the input variables temperature, time, and previous input levels, but these measures don't clarify the recent pandemic trends. In L.J. Kricka et al. (2020), mobility for measuring economic activities is used to capture the latest behavior. In this job, the predictive method uses ML algorithms. The analysis of the impact of COVID-19 on power and petroleum in China is conducted using a comparative regressive model and ANN model in L. Peng et al. (2020).
- *Assisting organizations*: AI resources are used in M. B. Schultz et al. (2020) to help organizations address their issues during the pandemic. In L. Tarrataca et al. (2020), the library resources and resource distribution during the pandemic were proposed to be optimized using an AI algorithm. The pandemic has made it impossible for the justice system to provide the service it needs. In systems like artificial lawyers, AI methods such as Ross intelligence, deep learning, and natural language processing are commonly used. The new problems have exacerbated the pressures to create intelligent systems to support the justice system. M. I. Abdelmageed et al. (2020) provide various ways for AI to assist in dealing with the issues of the pandemic.

An ongoing neural network to detect fraud during the pandemic is proposed in M. I. Uddin et al. (2020).

- *Help researchers*: This epidemic resulted in very crucial access to the latest scientific information. To examine this, the aim is to scan biological studies around COVID-19 (M. Moskal et al. 2020) in deep learning-based retrieval techniques that may be publically accessed for learning.
- *Education system*: After the pandemic, the education system was significantly disrupted, and many countries had used new educational platforms to support students. User satisfaction is very critical in these networks. An ANN is used in M. Simsek et al. (2020) to predict and forecast user satisfaction for education in China.
- *Economic impact management*: A data-guided dynamic clustering system has been proposed in M. A. Rahman et al. (2020) to mitigate the negative economic effects of COVID-19 flare-ups and alleviate lockdowns' economic impact. In view of the pandemic and economic and mobility aspects, the authors designed a model for localized lockdown through a clustering algorithm. Data-driven models for different countries used for epidemiological forecasts are described in M.-H. Tayarani-N. et al. (2021). The approach uses a thorough learning estimate of disease parameters to forecast cases and deaths and uses a genetic algorithm to best compromise between decision-making limitations and goals. At the end, a neural regressor is proposed in G. MacLaren et al. (2020) to clarify how COVID-19 affects Brazil.
- *Smart cities*: Several works developed ideas in smart city growth to address the difficulties caused by the pandemic for cities. In N. A. Brooks et al. (2020), AI approaches are used to assess the virus outbreaks, and methods for enhanced data exchange standardization protocols are proposed for smart city networks to improve pandemic management. In and between towns, human behavior changed drastically during the pandemic. A deep learning algorithm (N. Nawaz et al. 2020), combining strategic position sampling and a collection of lightweight invasive neural networks, is proposed to understand these patterns' changes. The model is generated for the recognition and calculation of specific elements in satellite images. A DNN is proposed in N. Norouzi et al. (2020) to forecast the impact of the pandemic on transport patterns.

4.6.5 For Pharmaceutical Studies

Finding an appropriate medicine will help decrease the disease mortality rate. Three major recurring choices, research therapy such as Remdesivir, and production of vaccines are the treatment methods for the disease. Repurposing medicines with few side effects for treating the disease is an effective and successful method in developing new therapeutic strategies. AI methods in pharmaceutical studies are used in some experiments in fighting COVID-19. Since the pandemic is ongoing, the AI opportunity should be exploited during the pharmaceutical testing and refurbishment phase (N. S. Punn et al. 2020).

- *Drug repurposing*: Combination therapy focused on drug repurposing is among the most common approaches. Medicines based on their mechanism, accompanied by dose-finding, are chosen to detect therapeutic synergy in multidrug treating. But it is a challenge to achieve this blend. To deal with this, 12 drug-doze parameters are proposed for an AI-based platform to identify therapeutics inhibiting lung cell infection. Predicting interactions in recommendation systems and drug development between heterogeneous graph-structured data applications and repurposing new disease medicines. In N. Soures et al. (2020), new drugs are discovered with ML algorithms. In P. Castorina et al. (2020), a model is created for repurposing drugs using the Naive Bayes algorithm. An AI-based algorithm is proposed in P. Khandelwal et al. (2020) to study 6,340 drugs to determine their anticipated SARS-CoV-2 efficacy. Finally, in R. Batra et al. (2020), an integral, network-based approach for identifying repurposes for this disease is proposed. The research presents a comprehensive knowledge graph covering 15 million borders across 39 types of drug, disease, protein, gene, pathway, and expression relationships published in several scientific journals. A deep learning network-based platform was used to identify 41 repurposable medicines. Clinical trials then validated the efficacy of the medicines. The authors argue that algorithms are not permitted to prescribe a particular drug, but that drug testing takes priority.
- *Finding potential medicines*: To uncover prospective treatment medicines for the disease, a library of 1,670 chemicals was established in a detailed study (R. F. Sear et al. 2020). In R. K. Pathan et al. (2020), a DNN was used to look for antivirals acting on the host target in experimental and approved drugs with possible disease activity. The algorithm investigates signatures of gene expression in close proximity to the SARS-CoV. In R. Magar et al. (2020), a transcriptional review is carried out to classify possible antiviral medicines derived from natural products or FDA-approved drugs using AI methods. An AI platform for identifying possible old antiviral drugs against COVID-19 established in R. Minetto et al. (2020) shows how Bayesian optimization can help prioritize candidates with equal computational power to improve the results further. A data-driven repository system is built in (R. Pandey et al. 2020) that incorporates ML and generates large-scale graphs to discover new medication candidates. A drug repositioning technique is proposed in S. Bandyopadhyay et al. (2020) to construct a model of learning prediction and find drugs to treat the disease.
- *Studying an immune system*: In another work (S. Cantürk et al. 2020), a neural network is carried out to detect possible therapeutic targets for COVID-19 repurpose drugs in the silicon analysis of the immune system protein interactome network with a single-cell RNA sequencing. In S. Ghamizi et al. (2020), the research aims to identify potential drugs that can block viral epitopes of the disease in finding peptides or antibody sequences.
- *Herbal medicines*: Certain herbal medicines are recommended to help cure the disease. MLP, SVM, and random forest algorithms are used to research

Indonesian herbal compounds and their drug efficacy. Furthermore, the 3D Virus mail protease structure-based method is used for a pharmacophore model approach in this method.

- *Analysis of the drug structure of the drug molecule*: The AI algorithm (S. Johnstone et al. 2020), which assesses the resemblance between the drug "progeny" and those parents that are already tested, is proposed to detect "progeny" drug "like" the "parents" tested for COVID-19. In [84], a group of 77 antiviral molecules, with their structural details, are analyzed using ML algorithms to classify possible therapies for crisis management. A deep learning algorithm is used in another work (S. Mahapatra et al. 2020) to classify molecular structures which may inhibit the virus. An *in vitro* therapy that demonstrates its efficacy is carried out by S. Mohanty et al. (2020). Reliable data on molecular interactions provide a basis for valuable data tools developed in drug protein–protein interaction networks. The analysis of these networks is based on a deep learning algorithm. The algorithm will predict unknown binding connections between medicinal products and human protein.
- *Study of existing drugs*: AI-oriented affinity provision for the classification of FDA drugs that can block coronavirus-entering cells by binding them to ACE2 or TMPRSS is proposed by S. Polyzos et al. (2020). Furthermore, a binding affinity prediction based on AI is suggested. In S. Ray et al. (2020), a pretrained, deep-learning drug–target interaction model is used to classify commercially available medicines, which can function on SARS viral proteins, known as the transformer molecules.
- *Studying techniques for drug discovery*: A systematic review of drug discovery techniques for COVID-19 based on the AI is suggested by S. Soni et al. (2020). In addition, S. Srinivasan et al. (2020) discussed how an IA-assisted prediction would contribute to the development of new disease medicines. An LSTM model is also trained to read the molecule's SMILES fingerprint and predict the molecule's IC50 when binding to RdRp (S.-W. Lee et al. 2020).
- *Molecular-design strategy*: In T. Chen et al. (2020), an AI algorithm was suggested to identify therapeutic biomolecules against COVID-19. The system uses a search algorithm for the Monte Carlo Tree and an ANN model of replacement for multitasking. In the framework (T. P. Mashamba-Thompson et al. 2020), an adaptive pretraining program combining the molecular autoencoder of SMILES and the directed sampling scheme of multiattributes is proposed. The approach uses guidance from latent-feature-qualified attribute predictors. This scheme produces new and optimized drug-like molecules for invisible viral targets with a protein-binding affinity predictor.
- *Virus sequence study*: The National Center for Biotechnology Information and Drug Virus Development Laboratory work together to establish a database model to find viral proteins and antiviral therapies that interact with them (T. P. Mashamba-Thompson et al. 2020). The model consists of virus protein sequences that serve as inputs and human antiviral medications that provide output.

- *Infection mechanism study*: ML-based models are paired with highly reliable combined docking simulations for rapid therapeutic molecule screening (T. Preethika et al. 2020). The test focuses on the affinities between the S-protein virus and the protein human interface in the host receiver region and the ACE2. The interaction of the host–virus can be limited or disrupted. The algorithm is used to find ligands that can be applied to two medicines. Studies show that pregnant women have health characteristics similar to nonpregnant women. An ML model was designed by T. Sundar et al. (2020) to predict the pregnancy safety profile of potential drugs based on proven medicinal sources with known pregnancy security.
- *Vaccine trials*: In the production of vaccines, ML methods have also been used. In the newly developed Vaxign-ML and Vaxign reverse vaccinology tool, ML algorithms are used to predict COVID-19 vaccine candidates (T. T. Nguyen et al. 2020). The AI algorithms in T. T. Nguyen et al. (2020) are used for studying the vaccine mutation conduct of the virus. It is alleged that the incidence of COVID-19 may be lowered by the vaccination of bacilli Calmette–Guerin (BCG). In V. Chenthamarakshan et al. (2020), the presence of such correlations is analyzed by ML algorithms. The authors used step-wise linear regression and k-means clustering.

4.7 BIG DATA FOR COVID-19

The nature of this virus is well understood. Big data technologies can store a large number of data on people with COVID-19 infections. The collected data can be further trained to develop potential preventive methods. This technology also allows storing all types of COVID-19-affected cases (infected, expired, and recovered). This expertise can effectively acknowledge the case and distribute resources for better public health protection (V. Lampos et al. 2020).

Big data offers a large amount of information and encourages scientists, healthcare professionals, and epidemiologists to make decisions to fight COVID-19. The virus can be monitored continuously worldwide and medical innovation developed (V. N. Ioannidis et al. 2020). In a particular area and population, the impact of COVID-19 can be predicted. Models for active feedback may be used at populations and demographic levels. It also helps to invent new methods of treatment. Big data can also provide potential outlets and tools and assist people in dealing with circumstances of stress. In general, this technology offers data to analyze transmission, movement, control, and disease prevention processes.

COVID-19 pandemic monitoring, control, testing, and prevention can be done by using big data analytics. It diversifies production and enhances the creation of vaccines with absolute expertise and deeper resources. Prevalent modeled data helps understand and give an advantage over the other method to predict the COVID-19 cure. Furthermore, big data offers information and analysis into how the infected individuals are put in containment. By collecting and implementing data with AI, China suppressed COVID-19, which led to a low spread rate. AI will play a significant role in biomedical science, natural language processing, social media, and scientific literature mining in many big data components of this pandemic (V. Sangiorgio et al. 2020).

4.7.1 Big Data Applications for COVID-19

This section contains essential COVID-19 big data applications.

1. *Identifying infected cases*: Owing to their capacity to store a large volume of data, big data can store all patients' complete medical history. This technology helps determine the contaminated cases and further risk analysis by supplying the collected data.
2. *History of travel*: Big data can analyze the danger of people's travel history. It helps recognize individuals who may be in contact with this virus's infected patient.
3. *Symptoms of fever*: Big data can retain the record of a patient's disease and other symptoms and indicate medical treatment. It contributes to the identification of suspicious cases and other misinformation with relevant details.
4. *Early identification of the virus*: Big data efficiently assists in the infected patient's early identification process. It helps analyze and classify individuals who might in the future be affected by this virus.
5. *Fast-moving disease identification and analysis:* Big data helps analyze the disease as quickly as possible. It can handle adequate disease knowledge.
6. *Lockdown information*: Big data can be used during lockdown to collect information on this virus. It can also control and track people's movement and the entire management of health.
7. *Persons entering or leaving the affected area*: Large numbers of people entering or leaving the affected area/city are analyzed by big data. These vast data allow health experts to identify the virus chances in these populations quickly.
8. *Faster medical care growth*: Big data can help quickly monitor the development of new drugs and equipment for current and potential medical needs. It spreads data and information about the virus and thus helps to improve over newly analyzed pandemics and epidemics.

4.7.2 Big Data for COVID-19 Pandemic: A Survey on the State of the Arts

Big data analytics for COVID-19 are mainly characterized by some key techniques from the big data definition, for example, the multidomain dataset analysis, deep analysis, high-dimensional analysis, and parallel computing. This section provides a summary of the state-of-the-art strategies enabled by big data (W. Alkady et al. 2020).

1. *Outbreak prediction*
   - In X. Zeng et al. (2020), it is proposed to use big datasettings of Italian Civil Protection sources to evaluate the outbreak potential of the disease.

In Wuhan, the first test was carried out to estimate the quarantine-contaminated population with COVID-19.

- In W. Xia et al. (2020), pandemic modeling is provided for interpreting the total numbers of infected people and the number of recovered cases in various regions such as Wuhan, Beijing, and Shanghai. Furthermore, in areas at high risk of a pandemic, this scheme will forecast the propensity of the COVID-19 outbreak.
- A comprehensive dataset of regions and countries like Korea and China is used as a basis in J. Zeng et al. (2020) for estimating the pandemic on a logistic model to assess the reliability of predictions.
- A large-scale data analysis approach from American cities is examined in (Y.-H. Jin et al. 2020) the United States. This method allows prediction errors to be calculated to improve the data modeling model for the estimate's precision.

2. *Virus spread tracking*
   - The big data analytical methods used by the National Health Commission of China to monitor the spread of COVID-19 are considered for using a large dataset of 854,424 persons. Analytical findings indicate that the positive cases are highly correlated with the population scale.
   - A big data analytical model is built with datasets from China, South Korea, Singapore, and Italy, and in Z. Allam et al. (2020), it is used to monitor the spread of viruses. This model predicts the maximum number of infected patients in a specific location.
   - It is claimed that the temperature-based model (Z. Allam et al. 2020) predicts how many cases of infection are correlated to a country's average temperature to monitor coronaviruses.
   - A big data unsupervised model, which distributed COVID-19 data gathered from Internet databases with the addition of a new metric for the number of news stories mentioning COVID-19 occurrences, was constructed in Z. Li et al. (2020).

3. *Coronavirus diagnosis/treatment*
   - The diagnosis of SARS-CoV-2 is proposed as a durable, responsive, precise, and very quantitative solution based on polymerase chain reactions. The proposed scheme has shown that the diagnosis of *Plasmodium falciparum* infections is an effective and inexpensive way.
   - In human cells that become infected with COVID-19 viruses, 6,381 proteins are suggested. This analysis aims to analyze data collected from the COVID-19 diagnosis storage in Kyoto Genes.
   - A series of large-scale clinical experiments, ranging from a normal and atypical manifestation of CT/X-ray imaging to hematological analysis and identification in the respiratory system, was introduced. These tests provide an extensive guide with valuable resources to diagnose and treat COVID-19.

4. *Vaccine/drug discovery*
   - A tool for investigating the SARS CoV, MERS CoV, and SARS-CoV-2 spike proteins and four earlier breakthrough human coronavirus strains is proposed. This permits a crucial vaccine production screening of SARS CoV-2 spike sequence and structure.
   - A large dataset from the National Centre of Biotechnology Information for Promoting the Manufacture of Vaccines will be used for the project. For the development of a new COVID-19 vaccine, several peptides were suggested.
   - A solution for drug research involving over 2,500 small molecules is which promotes drug repurposing against COVID-19.

4.8 CASE STUDY: HOW INDIA FIGHTS COVID-19 WITH AI AND BIG DATA

Long before the COVID-19 explosion, Bill Gates anticipated a nuclear war and an infectious virus as the next global disaster. The rest of the world, Wuhan's base, appears to have begun to move toward normality, but India is far from being "natural." In the last decade, studies showed that India is far behind from the WHO-prescribed patient–doctor ratio (1:1,456, according to the Economic Survey of India, 2019–20). For each doctor, that is, there are 456 additional patients.

In the 2021 Union budget, the health infrastructure in the country has been improved by approximately 64,000 crore rupees. Until last year, only 1.9% of India's healthcare industry accounted for Artificial Intelligence (AI). We now face a dire need to introduce technology into the health technology environment of the country more than ever before.

Companies use ML software as a service and data analytics, etc. to tackle the current pandemic.

1. *Monitoring:*Mask violators are detected through Madurai and Telangana's AI technology. Video applications parse public CCTVS data streams to detect people who violate protocols of the Health Department. The Madurai City Police had arrested nearly 47,000 people by November and fined nearly Rs 89 lakh. To alert commercial stores and offices of social breaches during the pandemic, Pune-based start-up Glimpse Analytics has been using AI.
2. *Diagnosis:* In some cases, new COVID mutants, leading to testing centers with overflowing CT scans and chest X-rays, were not detected in the tests for RT-PCR to validate the diagnosis. DRDO had developed AI-based smart COVID detection application software ATMAN AI at its AI and Robotics Centre. This software performs Image classification under "normal," "COVID-19," and "pneumonia" using chest X-rays. The Deep Convolutional Neural Network is assisted by ATMAN AI. Beyond transmitting the images to the neural net, the program analyzes the images until it handles the different illuminations of the X-Ray images. You may use a smartphone, tablet, and laptop or computer to access ATMAN AI. According to DRDO, ATMAN AI showed 96.73% accuracy in RT-PCR-positive patients with

automated chest X-rays. So far, doctors at the HCG Center for Academics and Research and Ankh Life Care have checked and validated ATMAN AI in Bengaluru. IIT Kharagpur also introduced COVIRAP in April, a diagnostic technique for detecting infectious illnesses like COVID-19. The media reports state that COVIRAP consists of a preprogrammative monitoring unit, a genomic analysis special detection unit, and a custom smartphone app for displaying the test results. Mumbai start-up Qure.ai has built qXR, a platform for monitoring and diagnosing COVID-19 based on deep learning. Tata Consultancy Services have used AI for X-ray plate screening and coronavirus diagnostics.

3. *Assistance:* Last-month, Yellow Messenger, a conversational CX platform, launched Yellow Messenger Cares that empowers clinics, hospitals, and companies. If you are a company that contributes to COVID-19, we will support you by sharing the load. Cofounder Raghu Ravinutala mentioned one might be an NGO, a healthcare center, an insurance provider, or just a voluntary organization. All one has to do is let them know the kind of chatbot services one are contributing to and how they will help one start a chat bot.
4. *Information:* Accenture and Microsoft partnered with Digital India Corporation of the Indian Government to equip its people with the AI chatbot MyGov Saathi to supply reliable, useful, and recent COVID-19 details. Accenture mentioned that Saathi blends NLP and AI data analytics and conversation to support 50,000 users a day. Likewise, WhatsApp-based Introbot offers users an advanced and checked bed database, oxygen cylinders, plasma, and other medical devices on demand. The AI Community Manager informed about more than 300 cities worldwide and responded in the first week of its launch to more than five lakhs and COVID-19 victims.
5. *Drug repurposing:* The drug discovery process is not only time-consuming and risky but also very costly. In this respect, AI is used to repurpose medications to treat COVID-19. For example, hydroxychloroquine (used for the treatment of malaria) and Remdesivir (Ebola Drug) against COVID-19 were assessed by the Indo-German organization Innoplexus. Last year, Delhi's Indraprastha Institute of Information Technology created an AI model for the repurposing of drugs. First, the AI model calculates the similarity between the drug's chemical structure and the virus genomic structure. Then, it checks and picks the drug that has successfully treated viruses with a COVID-19 genomics framework, using the historical efficacy of the drug.

4.9 CONCLUSION

The novel coronavirus challenges the lay person in various ways, and it is like an almost unbeatable enemy for healthcare professionals, medical researchers, and policymakers. There is an expectation that AI and other emerging technology can reduce the gravity of the problems lying ahead of us. But we cannot go ahead—it goes without saying, and time and patience will be needed for an effective long-term response to the virus. AI is not a magic bullet, and its existing capabilities are practical and logical. Nevertheless, it has undoubtedly helped us understand better what we are dealing with and how we can find a solution to this pandemic. AI and big data have already laid the groundwork for reducing potential transmissions—and for us to find a correct solution toward eradicating the virus.

REFERENCES

Abdelmageed, M. I., A. H. Abdelmoneim, M. I. Mustafa, N. M. Elfadol, N. S. Murshed, S. W. Shantier, A. M. Makhawi, "Design of a multiepitope-based peptide vaccine against the E protein of human COVID-19: An immunoinformatics approach," *BioMed Research International* (2020).

Abdollahi, A., M. Rahbaralam, "Effect of temperature on the transmission of COVID-19: A machine learning case study in Spain," *medRxiv* (2020).

Abdulla, A., B. Wang, F. Qian, T. Kee, A. Blasiak, Y. H. Ong, L. Hooi, F. Parekh, R. Soriano, G. G. Olinger, et al., "Project IDentif.AI: Harnessing artificial intelligence to rapidly optimize combination therapy development for infectious disease intervention," *Advanced Therapeutics* 3(7) (2020) 2000034.

Alimadadi, A., S. Aryal, I. Manandhar, P. B. Munroe, B. Joe, X. Cheng, "Artificial intelligence and machine learning to fight COVID-19," (2020).

Alkady, W., M. Zanaty, H. M. Afify, "Computational predictions for protein sequences of COVID-19 virus via machine learning algorithms" (2020).

Allam, Z., G. Dey, D. S. Jones, "Artificial intelligence (AI) provided early detection of the coronavirus (COVID-19) in China and will influence future urban health policy internationally," *AI* 1 (2) (2020) 156–165.

Anshari, M., M. N. Almunawar, S. A. Lim, A. Al-Mudimigh, "Customer relationship management and big data enabled: Personalisation & customisation of services," *Applied Computing and Informatics* 15(2) (2019) 94–101.

Arora, K., A. S. Bist, "Artificial intelligence based drug discovery techniques for COVID-19 detection," *Aptisi Transactions on Technopreneurship (ATT)* 2(2) (2020) 120–126.

Avchaciov, K., O. Burmistrova, P. Fedichev, "AI for the repurposing of approved or investigational drugs against COVID-19," (2020). https://doi.org/10.13140/RG.2.2.20588.10886.

Bandyopadhyay, S., S. Dutta, "Detection of fraud transactions using recurrent neural network during COVID-19," (2020).

Banerjee, A., D. Santra, S. Maiti, "Energetics based epitope screening in SARS CoV-2 (COVID 19) spike glycoprotein by immuno-informatic analysis aiming to a suitable vaccine development," *bioRxiv* (2020).

Batra, R. H. Chan, G. Kamath, R. Ramprasad, M. J. Cherukara, S. Sankaranarayanan, "Screening of therapeutic agents for COVID-19 using machine learning and ensemble docking simulations," *arXiv:2004.03766* (2020).

Beck, B. R., B. Shin, Y. Choi, S. Park, K. Kang, "Predicting commercially available antiviral drugs that may act on the novel coronavirus (sars-cov-2) through a drug-target interaction deep learning model," *Computational and Structural Biotechnology Journal* 18 (2020) 784–790.

Bonacini, L., G. Gallo, F. Patriarca, "Drawing policy suggestions tonight COVID-19 from hardly reliable data. A machine-learning contribution on lockdowns analysis," *Technical Report, GLO Discussion Paper* (2020).

Brooks, N. A., A. Puri, S. Garg, S. Nag, J. Corbo, A. El Turabi, N. Kaka, R. W. Zemmel, P. K. Hegarty, A. M. Kamat, "COVID-19 mortality and BCG vaccination: Defining the link using machine learning," (2020).

Cantürk, S., A. Singh, P. St-Amant, J. Behrmann, "Machine learning driven drug repurposing for COVID-19," *arXiv:2006.14707* (2020).

Castorina, P., A. Iorio, D. Lanteri, "Data analysis on coronavirus spreading by macroscopic growth laws," *arXiv:2003.00507* (2020).

Chandra, G., R. Gupta, N. Agarwal, "Role of artificial intelligence in transforming the justice delivery system in COVID-19 pandemic," (2020) 344–350.

Chen, T., L. Peng, X. Yin, J. Rong, J. Yang, G. Cong, "Analysis of user satisfaction with online education platforms in China during the COVID-19 pandemic," *Healthcare* 8 (2020) 200 Multidisciplinary Digital Publishing Institute.

Chenthamarakshan, V., P. Das, S. C. Hoffman, H. Strobelt, I. Padhi, K. W. Lim, B. Hoover, M. Manica, J. Born, T. Laino, A. Mojsilovic, "Cogmol: Target-specific and selective drug design for COVID-19 using deep generative models," *arXiv:2004.01215* (2020).

Dule, C. S., R. KM, M. DH, "Challenges of artificial intelligence to combat COVID-19," *SSRN* 3608764 (2020).

Erlina, L., R. I. Paramita, W. A. Kusuma, F. Fadilah, A. Tedjo, I. P. Pratomo, N. S. Ramadhanti, A. K. Nasution, F. K. Surado, A. Fitriawan, et al., "Virtual screening on Indonesian herbal compounds as COVID-19 supportive therapy: Machine learning and pharmacophore modeling approaches," (2020).

Ghamizi, S., R. Rwemalika, L. Veiber, M. Cordy, T. F. Bissyande, M. Papadakis, J. Klein, Y. L. Traon, "Data-driven simulation and optimization for COVID-19 exit strategies," *arXiv:2006.07087* (2020).

Giordano, G., F. Blanchini, R. Bruno, P. Colaneri, A. Di Filippo, A. Di Matteo, M. Colaneri, "Modelling the COVID-19 epidemic and implementation of population-wide interventions in Italy," *Nature Medicine* (2020) 1–6.

Goh, G., A. K. Dunker, J. Foster, V. Uversky, "A novel strategy for the development of vaccines for sars-cov-2 (COVID-19) and other viruses using ai and viral shell disorder," (2020).

Gysi, D. M., Ítalo Do Valle, M. Zitnik, A. Ameli, X. Gan, O. Varol, S. D. Ghiassian, J. J. Patten, R. Davey, J. Loscalzo, Albert-László Barabási, "Network medicine framework for identifying drug repurposing opportunities for COVID-19," *arXiv:2004.07229* (2020).

Haleem, A., M. Javaid, I. H. Khan, R. Vaishya, "Significant applications of big data in COVID-19 pandemic," *Indian Journal of Orthopaedics* 54 (2020) 526–528. https://doi.org/10.1007/s43465-020-00129-z

Haleem, A., M. Javaid, R. Vaishya, "Effects of COVID 19 pandemic in daily life," *Current Medicine Research and Practice* (2020). https://doi.org/10.1016/j.cmrp.2020.03.011

Hashem, I. A. T., A. E. Ezugwu, M. A. Al-Garadi, I. N. Abdullahi, O. Otegbeye, Q. O. Ahman, G. C. E. Mbah, A. K. Shukla, H. Chiroma, "A machine learning solution framework for combatting COVID-19 in smart cities from multiple dimensions," *medRxiv* (2020). https://doi.org/10.1101/2020.05.18.20105577

Heiser, K., P. F. McLean, C. T. Davis, B. Fogelson, H. B. Gordon, P. Jacobson, B. Hurst, B. Miller, R. W. Alfa, B. A. Earnshaw, M. L. Victors, Y. T. Chong, I. S. Haque, A. S. Low, C. C. Gibson, "Identification of potential treatments for COVID-19 through artificial intelligence-enabled phenomic analysis of human cells infected with sars-cov-2," *bioRxiv* (2020).

Ioannidis, V. N., D. Zheng, G. Karypis, "Few-shot link prediction via graph neural networks for COVID-19 drug-repurposing," *arXiv:2007.10261* (2020).

Islam, A. N., S. Laato, S. Talukder, E. Sutinen, "Misinformation sharing and social media fatigue during COVID-19: An affordance and cognitive load perspective," *Technological Forecasting and Social Change* 159 (2020) 120201.

Ivanov, D., A. Dolgui, "Viability of intertwined supply networks: Extending the supply chain resilience angles towards survivability. A position paper motivated by COVID-19 outbreak," *International Journal of Production Research* 58(10) (2020) 2904–2915.

Jin, Y.-H., L. Cai, Z.-S. Cheng, H. Cheng, T. Deng, Y.-P. Fan, C. Fang, D. Huang, L.-Q. Huang, Q. Huang et al., "A rapid advice guideline for the diagnosis and treatment of 2019 novel coronavirus (2019-ncov) infected pneumonia (standard version)," *Military Medical Research* 7(1) (2020) 4.

Johnstone, S., "A viral warning for change. COVID-19 versus the Red Cross: Better solutions via blockchain and artificial intelligence," *University of Hong Kong Faculty of Law Research Paper* (2020/005) (February 3, 2020).

Khandelwal, P., A. Khandelwal, S. Agarwal, D. Thomas, N. Xavier, A. Raghuraman, "Using computer vision to enhance safety of workforce in manufacturing in a post COVID world," *arXiv:2005.05287* (2020).

Kim, J., Y. Cha, S. Kolitz, J. Funt, R. Escalante Chong, S. Barrett, B. Zeskind, R. Kusko, H. Kaufman, et al., "Advanced bioinformatics rapidly identifies existing therapeutics for patients with coronavirus disease-2019 (COVID-19)," (2020).

Kricka, L. J., S. Polevikov, J. Y. Park, P. Fortina, S. Bernardini, D. Satchkov, V. Kolesov, M. Grishkov, "Artificial intelligence-powered search tools and resources in the fight against COVID-19," *EJIFCC* 31(2) (2020) 106.

Lampos, V., S. Moura, E. Yom-Tov, I. J. Cox, R. McKendry, and M. Edelstein, "Tracking COVID-19 using online search," *arXiv:2003.08086* (2020).

Lee, S.-W., H. Jung, S. Ko, S. Kim, H. Kim, K. Doh, H. Park, J. Yeo, S.-H. Ok, J. Lee, S. Choi, S. Hwang, E.-Y. Park, G.-J. Ma, S.-J. Han, K.-S. Cha, N. Sung, J.-W. Ha, "Carecall: A call-based active monitoring dialog agent for managing COVID-19 pandemic," *arXiv:2007.02642* (2020).

Li, C., D. N. Debruyne, J. Spencer, V. Kapoor, L. Y. Liu, B. Zhou, L. Lee, R. Feigelman, G. Burdon, J. Liu et al., "High sensitivity detection of coronavirus SARS-CoV-2 using multiplex PCR and a multiplex-PCR based metagenomic method," *bioRxiv* (2020).

Li, Z., X. Li, Y.-Y. Huang, Y. Wu, L. Zhou, R. Liu, D. Wu, L. Zhang, H. Liu, X. Xu et al., "FEP-based screening prompts drug repositioning against COVID-19," *bioRxiv* (2020).

López-Cortés, A., P. Guevara-Ramírez, N. C. Kyriakidis, C. Barba-Ostria, Á. L. Cáceres, S. Guerrero, C. R. Munteanu, E. Tejera, E. Ortiz-Prado, D. Cevallos-Robalino, et al., "In silico analyses of immune system protein interactome network, single-cell RNA sequencing of human tissues, and artificial neural networks reveal potential therapeutic targets for drug repurposing against COVID-19," (2020).

MacLaren, G., D. Fisher, D. Brodie, "Preparing for the most critically ill patients with COVID-19: The potential role of extracorporeal membrane oxygenation," *JAMA* (2020). https://doi.org/10.1001/jama.2020.2342.

Magar, R., P. Yadav, A. B. Farimani, "Potential neutralizing antibodies discovered for novel corona virus using machine learning," *bioRxiv* (2020).

Mahapatra, S., P. Nath, M. Chatterjee, N. Das, D. Kalita, P. Roy, S. Satapathi, "Repurposing therapeutics for COVID-19: Rapid prediction of commercially available drugs through machine learning and docking," *medRxiv* (2020).

Mashamba-Thompson, T. P., E. D. Crayton, "Blockchain and artificial intelligence technology for novel coronavirus disease-19 self-testing," (2020).

Minetto, R., M. P. Segundo, G. Rotich, S. Sarkar, "Measuring human and economic activity from satellite imagery to support city scale decision-making during COVID-19 pandemic," *arXiv:2004.07438* (2020).

Mohanty, S., M. Harun AI Rashid, M. Mridul, C. Mohanty, S. Swayamsiddha, "Application of artificial intelligence in COVID-19 drug repurposing," *Diabetes & Metabolic Syndrome: Clinical Research & Reviews* 14(5) (2020) 1027–1031.

Moskal, M., W. Beker, R. Roszak, E. P. Gajewska, A. Wołos, K. Molga, S. Szymkuć, G. Grynkiewicz, B. Grzybowski, "Suggestions for second-pass anti-COVID-19 drugs based on the artificial intelligence measures of molecular similarity, shape and pharmacophore distribution," (April 2020).

Mowery, C. T., A. Marson, Y. S. Song, C. J. Ye, "Improved COVID-19 serology test performance by integrating multiple lateral flow assays using machine learning," *medRxiv* (2020).

Nawaz, N., A. M. Gomes, M. A. Saldeen, "Artificial intelligence (AI) applications for library services and resources in COVID-19 pandemic," *Artificial Intelligence (AI)* 7(18) (2020).

Nguyen, T. T., M. Abdelrazek, D. T. Nguyen, S. Aryal, D. T. Nguyen, A. Khatami, "Origin of novel coronavirus (COVID-19): A computational biology study using artificial intelligence," *bioRxiv* (2020).

Norouzi, N., G. Zarazua de Rubens, S. Choupanpiesheh, P. Enevoldsen, "When pandemics impact economies and climate change: Exploring the impacts of COVID-19 on oil and electricity demand in China," *Energy Research & Social Science* 68 (2020) 101654.

Ong, E., M. U. Wong, A. Huffman, Y. He, "COVID-19 coronavirus vaccine design using reverse vaccinology and machine learning," *bioRxiv* (2020).

Pandey, R., V. Gautam, C. Jain, P. Syal, H. Sharma, K. Bhagat, R. Pal, L. S. Dhingra, Arushi, L. Patel, M. Agarwal, S. Agrawal, M. Arora, B. Rana, P. Kumaraguru, T. Sethi, "A machine learning application for raising wash awareness in the times of COVID-19 pandemic," *arXiv:2003.07074* (2020).

Pathan, R. K., M. Biswas, M. U. Khandaker, "Time series prediction of COVID-19 by mutation rate analysis using recurrent neural network based lstm model," *Chaos, Solitons & Fractals* 138 (2020) 110018.

Peng, L., W. Yang, D. Zhang, C. Zhuge, L. Hong, "Epidemic analysis of COVID-19 in China by dynamical modeling," *arXiv:2002.06563* (2020).

Pham, Q. V., D. C. Nguyen, T. Huynh-The, W. J. Hwang, P. N. Pathirana, "Artificial Intelligence (AI) and big data for coronavirus (COVID-19) pandemic: A survey on the state-of-the-arts," *IEEE Access* 8 (2020) 130820–130839. https://doi.org/10.1109/ACCESS.2020.3009328.

Pirouz, B., H. JavadiNejad, G. Violini, B. Pirouz, "The role of swab tests to decrease the stress by COVID-19 on the health system using AI, MLR & statistical analysis," *medRxiv* (2020).

Pokkuluri, K. S., S. U. Devi Nedunuri, "A novel cellular automata classifier for COVID-19 prediction," *Journal of Health Sciences* 10(1) (2020) 34–38.

Polyzos, S., A. Samitas, A. E. Spyridou, "Tourism demand and the COVID-19 pandemic: An LSTM approach," *Tourism Recreation Research* (2020) 1–13.

Prakash, A., S. Muthya, T. P. Arokiaswamy, R. S. Nair, "Using machine learning to assess COVID-19 risks," *medRxiv* (2020).

Preethika, T., P. Vaishnavi, J. Agnishwar, K. Padmanathan, S. Umashankar, S. Annapoorani, M. Subash, K. Aruloli, "Artificial intelligence and drones to combat COVID-19," (2020).

Punn, N. S., S. K. Sonbhadra, S. Agarwal, "Monitoring COVID-19 social distancing with person detection and tracking via fine-tuned YOLO v3 and deepsort techniques," *arXiv:2005.01385* (2020).

Qin, B., D. Li, "Identifying facemask-wearing condition using imagesuper-resolution with classification network to prevent COVID-19," (2020).

Rahman, M. A., N. Zaman, A. T. Asyhari, F. Al-Turjman, M. Z. AlamBhuiyan, M. Zolkipli, "Data-driven dynamic clustering framework for mitigating the adverse economic impact of COVID-19 lockdown practices," *Sustainable Cities and Society* 62 (2020) 102372.

Randhawa, G. S., M. P. Soltysiak, H. El Roz, C. P. de Souza, K. A. Hill, L. Kari, "Machine learning using intrinsic genomic signatures for rapid classification of novel pathogens: COVID-19 case study," *PLoS ONE* 15(4) (2020) e0232391.

Rao, A. S. S., J. A. Vazquez, "Identification of COVID-19 can be quicker through artificial intelligence framework using a mobile phone-based survey when cities and towns are under quarantine," *Infection Control & Hospital Epidemiology* 41(7) (2020) 826–830.

Ray, S., S. Lall, A. Mukhopadhyay, S. Bandyopadhyay, A. Schönhuth, "Predicting potential drug targets and repurposable drugs for COVID-19 via a deep generative model for graphs," *arXiv:2007.02338* (2020).

Riccardi, A., J. Gemignani, F. Fernandez-Navarro, A. Heffernan, "Optimisation of non-pharmaceutical measures in COVID-19 growth via neural networks," (2020).

Sangiorgio, V., F. Parisi, "A multicriteria approach for risk assessment of COVID-19 in urban district lockdown," *Safety Science* 130 (2020) 104862.

Schultz, M. B., D. Vera, D. A. Sinclair, "Can artificial intelligence identify effective COVID-19 therapies?," *EMBO Molecular Medicine* (2020) e12817.

Sear, R. F., N. VelÃsquez, R. Leahy, N. J. Restrepo, S. E. Oud, N. Gabriel, Y. Lupu, N. F. Johnson, "Quantifying COVID-19 content in the online health opinion war using machine learning," *IEEE Access* 8 (2020) 91886–91893.

SesagiriRaamkumar, A., S. G. Tan, H. L. Wee, "Use of health belief model-based deep learning classifiers for COVID-19 social media content to examine public perceptions of physical distancing: Model development and case study," *JMIR Public Health Surveill* 6(3) (2020) e20493.

Shtar, G., L. Rokach, B. Shapira, E. Kohn, M. Berkovitch, M. Berlin, "Treating COVID-19 during pregnancy: Using artificial intelligence to evaluate medication safety," (2020).

Siau, K., R. Lian, "Artificial intelligence in COVID-19 pandemic management and control," (2020).

Simsek, M., B. Kantarci, "Artificial intelligence-empowered mobilization of assessments in COVID-19-like pandemics: A case study for early flattening of the curve," *International Journal of Environmental Research and Public Health* 17(10) (2020) 3437.

Soni, S., K. Roberts, "An evaluation of two commercial deep learning based information retrieval systems for COVID-19 literature," *arXiv:2007.03106* (2020).

Sonntag, D., "AI in medicine, COVID-19 and springer nature's open access agreement," *Kunstliche Intelligenz* 34(2) (2020) 123.

Soures, N., D. Chambers, Z. Carmichael, A. Daram, D. P. Shah, K. Clark, L. Potter, D. Kudithipudi, "Sirnet: Understanding social distancing measures with hybrid neural network model for COVID-19 infectious spread," *arXiv:2004.10376* (2020).

Srinivasan, S., R. Batra, H. Chan, G. Kamath, M. J. Cherukara, S. Sankaranarayanan, "Artificial intelligence guided de novo molecular design targeting COVID-19," (June 2020).

Stebbing, J., V. Krishnan, S. de Bono, S. Ottaviani, G. Casalini, P. J. Richardson, V. Monteil, V. M. Lauschke, A. Mirazimi, S. Youhanna, et al., "Mechanism of baricitinib supports artificial intelligence predicted testing in COVID-19 patients," *EMBO Molecular Medicine* (2020).

Sundar, T., K. Menaka, G. Vinotha, "Artificial intelligence suggested repositionable therapeutics for managing COVID-19: An investigation with machine learning algorithms and molecular structures," (2020).

Tarrataca, L., C. M. Dias, D. B. Haddad, E. F. Arruda, "Flattening the curves: On-off lockdown strategies for COVID-19 with an application to Brazil," *arXiv:2004.06916* (2020).

Tátrai, D., Z. Várallyay, "COVID-19 epidemic outcome predictions based on logistic fitting and estimation of its reliability," *arXiv:2003.14160* (2020).

Tayarani-N, M.-H., "Applications of artificial intelligence in battling against COVID-19: A literature review," *Chaos, Solitons and Fractals* 142 (2021) 1–31.

Uddin, M. I., S. A. A. Shah, M. A. Al-Khasawneh, "A novel deep convolutional neural network model to monitor people following guidelines to avoid COVID-19," *Journal of Sensors* 2020 (2020).

Wang, B., Y. Sun, T. Q. Duong, L. D. Nguyen, L. Hanzo, "Risk-aware identification of highly suspected COVID-19 cases in social IOT: A joint graph theory and reinforcement learning approach," *IEEE Access* 8 (2020) 115655–115661.

Wang, D., F. Zuo, J. Gao, Y. He, Z. Bian, S. Duran, C. N. Bernardes, J. Wang, J. Petinos, K. Ozbay, et al., "Agent-based simulation model and deep learning techniques to evaluate and predict transportation trends around COVID-19" (2020).

Wen, A., L. Wang, H. He, S. Liu, S. Fu, S. Sohn, J. A. Kugel, V. C. Kaggal, M. Huang, Y. Wang, F. Shen, J. Fan, H. Liu, "An aberration detection-based approach for sentinel syndromic surveillance of COVID-19 and other novel influenza-like illnesses," *medRxiv* (2020). https://doi.org/10.1101/2020.06.08.20124990.

Wu, J., "How artificial intelligence can help fight coronavirus," *Forbes* (2020). [Online]. Available at: www.forbes.com/sites/cognitiveworld/2020/03/19/how-artificial-intelligence-can-help-fight-coronavirus/?sh=7bb4312f4d3a. [Accessed: 01-Apr-2021].

Xia, W., T. Sanyi, C. Yong, F. Xiaomei, X. Yanni, X. Zongben, "When will be the resumption of work in Wuhan and its surrounding areas during COVID-19 epidemic? A data-driven network modelling analysis," *SCIENTIA SINICA Mathematica* (2020). https://doi.org/10.1360/SSM-2020–0037.

Yang, D., E. Yurtsever, V. Renganathan, K. A. Redmill, ÜmitÖzgüner, "A vision-based social distancing and critical density detection system for COVID-19," *arXiv:2007.03578* (2020).

Zeng, X., X. Song, T. Ma, X. Pan, Y. Zhou, Y. Hou, Z. Zhang, G. Karypis, F. Cheng, "Repurpose open data to discover therapeutics for COVID-19 using deep learning," *arXiv:2005.10831* (2020).

Zhavoronkov, A., B. Zagribelnyy, A. Zhebrak, V. Aladinskiy, V. Terentiev, Q. Vanhaelen, D. Bezrukov, D. Polykovskiy, R. Shayakhmetov, A. Filimonov, M. Bishop, S. McCloskey, E. Leija, D. Bright, K. Funakawa, Y.-C. Lin, S.-H. Huang, H.-J. Liao, A. Aliper, Y. Ivanenkov, "Potential non-covalent SARS-COV-2 3c-like protease inhibitors designed using generative deep learning approaches and reviewed by human medicinal chemist in virtual reality," (2020).

Zhu, J., Y.-Q. Deng, X. Wang, X.-F. Li, N.-N. Zhang, Z. Liu, B. Zhang, C.-F. Qin, Z. Xie, "An artificial intelligence system reveals liquiritin inhibits sars-cov-2 by mimicking type I interferon," *bioRxiv* (2020). https://doi.org/10.1101/2020.05.02.074021.

5 Blockchain in Artificial Intelligence

*Aditya Singh, Aayush Saxena, Ramani S, and *Marimuthu Karuppiah*

CONTENTS

DOI: 10.1201/9781003152392-5

5.1 INTRODUCTION

It is verifiable that Artificial Intelligence (AI) and blockchain ideas are spreading at a remarkable rate. The two advancements have a particular level of mechanical intricacy and multidimensional business suggestions. Now, a typical misunderstanding idea about blockchain, specifically, is that blockchain is decentralized and isn't constrained by anybody. In any case, the hidden improvement of a blockchain framework is as yet credited to a bunch of center designers. Accept keen agreement, for instance, it is basically an assortment of codes (or capacities) and information (or states) that are customized and conveyed on a blockchain (say, Ethereum) by various human software engineers. It is in this way, lamentably, less inclined to be liberated from provisos and blemishes (Angraal, S. et al. 2017). In this chapter, through a concise outline about how man-made reasoning could be utilized to convey without bug keen agreement to accomplish the objective of blockchain 2.0, we to underscore that the blockchain execution can be helped or upgraded by means of different AI strategies. The collusion of AI and blockchain is required to make various prospects (Bagga, P. et al. 2021).

Blockchain and AI have acquired the maximum amount of exploration consideration in this last decade. Blockchain is an appropriate record which is dependent on advanced records which are shared by an organization consisting of many members. Blockchain innovation has an expected limit in numerous areas like international payments, secure information sharing and advertising, and store network executives. Whereas, AI is utilized for building up of the machines/software equipped for executing undertakings that require insight (Benchoufi, M. et al. 2017).

Blockchain and AI have propelled advances and growth that have resulted in development across almost every industry. AI alludes to machines that are worked to perform shrewd errands that have generally been refined by people. Blockchain is a decentralized organization of PCs that records and stores information to show an ordered arrangement of occasions on a straightforward and permanent record framework. AI and blockchain are ending up being an incredible amazing pair, improving pretty much each field of the various industries where it is applied or implemented. Blockchain and AI combine to redesign everything including the food store network coordination along with medical services sharing their records as well as media sovereignties along with monetary security. Blockchain–AI assembly is inescapable on the grounds that both are arrangements with information and worth. Blockchain empowers secure capacity and sharing of information or anything of significant worth. AI can examine and create bits of knowledge from information to produce esteem (Bera, B. et al. 2020).

Blockchain is a disruptive innovation that empowers the improvement of dependable applications, without the requirement of trust between network peers. Blockchain innovation makes worldwide and permanent archives that ensure nonrenouncement and responsibility of putting away data. Furthermore, the blast in the age and accessibility of information on PC networks raises the test of handling and overseeing a lot of information at any point lower latencies. As an outcome, man-made brainpower and AI strategies experience huge upgrades and arise as empowering innovations for the cutting-edge organizations. This uncommon version is committed to these new innovations that shape the world to have more dependable PC organizations while empowering new dispersed and information-driven security applications and administrations (Chen, X. et al. 2018).

The Internet of Things (IoT) has been industrializing in a few true applications in recent years, for example, smart commuting/transportation and smart cities, in order to make the human life as stable as possible. Due to the growing technological advances of IoT, a large quantity of detecting data/records are being produced from a variety of sensors gadgets in the IoT industry. AI plays an important role in breaking down large amounts of data by acting as a solid logical instrument that continuously does a flexible and precise investigation of data. Notwithstanding, the plan and improvement of a helpful large information examination instrument utilizing AI have a few difficulties, like unified engineering, security, protection, asset requirements, and the absence of enough preparing information. Whereas, as an upcoming innovative idea, blockchain upholds a localized design. It provides a safe-sharing of information along with assets to all the different hubs of the IoT technology which is urged to eliminate brought together control and has the ability to defeat the current difficulties in AI. One of the primary objectives of this examination is to plan and build up an IoT design with blockchain and AI to help a compelling enormous information investigation. In this research paper, we proposed a blockchain which is empowered by Intelligent IoT' Architecture along with AI which gives an effective method of combining all three technologies—AI, blockchain along with IoT with the present status of the craftsmanship procedures, and applied methods (Cong, L. W. et al. 2019).

5.1.1 Difference between Blockchain and AI

To begin with, blockchain has a number of stability, versatility, and proficiency issues. Reasonability, reliability, and security are all problems that AI faces. If these two technologies are combined, the next computerized era will emerge.

The argument here is that blockchain provides AI with trustlessness, stability, and reasonableness. While AI contributes its expertise to the development of AI systems based on blockchain to achieve adaptability and which can be used fully for personalization and administration (Cong, L. W. et al. 2019).

5.1.2 Blockchain for AI (Classification and Protection) and AI for Blockchain (Security and Straightforwardness)

Blockchain can allow decentralized commercial centers and collaboration stages that can be used for a variety of AI applications such as power data and calculations. These would pave the way for a slew of new technologies and the expanded use of AI. Safe information sharing—Because AI involves overseeing large amounts of data that is used to prepare computers, there is a need for a more effective and secure method of sharing data in the form of a ballot and stored. Furthermore, one of the basic and growth factors is confidentiality, which involves overseeing massive knowledge gaps and misuse of individual data. Figure 5.1 shows opportunities brought by AI to address challenges of blockchain.

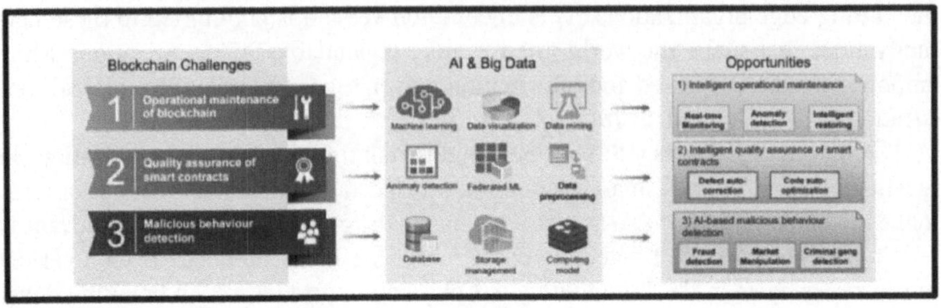

FIGURE 5.1 Opportunities brought by artificial intelligence to address challenges of blockchain.

Your data is valuable to you. More than just exchanging and managing client information, blockchain technology allows for the sale of information through smart contracts, removing the need for middlemen and making transactions safer and more private. This commercial center lowers the bar, allowing smaller companies to participate. Selling your extra computational power—blockchain will allow better appropriation of calculating power, which is essential for AI, and preparation in AI, through establishing a decentralized market for selling calculating power, is referred to as blockchain-based distributed computing. GPUs are only used for a small portion of the time; this unused processing time can be used to give AI-savvy contracts and be compensated).

5.1.3 Blockchain and AI: A Great Match

Blockchain and AI are the two most abundantly debated developments in today's technological advancements despite the fact that the parties creating each technology are diametrically opposed. According to PwC, AI-enabled technological developments would add \$15.7 trillion to the global economy, resulting in a 14% rise in global GDP. According to Gartner, the market value contributed through blockchain technological developments will reach of \$3.1 trillion in the same year (Croman, K. et al. 2016).

Blockchain is a spread out, decentralized, and nonchangeable ledger that is used in order to store some of the encrypted data, according to the given description. AI, on the other hand, is the driver or, to put it another way, a human-made "brain" that can assist us in data analytics and decision-making. We should both agree on one thing: despite the fact that all technologies have their own level of sophistication, both AI and blockchain can support and learn from each other in this collaboration. Combining these technologies makes sense because they can influence and enact data in various ways, and it can take data manipulation to great heights. While doing this in parallel, we can also integrate ML and AI with blockchain and vice versa, which would boost the fundamental architecture of blockchain while also increasing the ability of AI. Furthermore, blockchain will help us track and understand why machine learning (ML) decisions are made, making AI more coherent and intuitive. The blockchain and its ledger will store all of the data and variables that go into ML decision (Yaga, D. et al. 2018). Figure 5.2 shows applications of AI and blockchain.

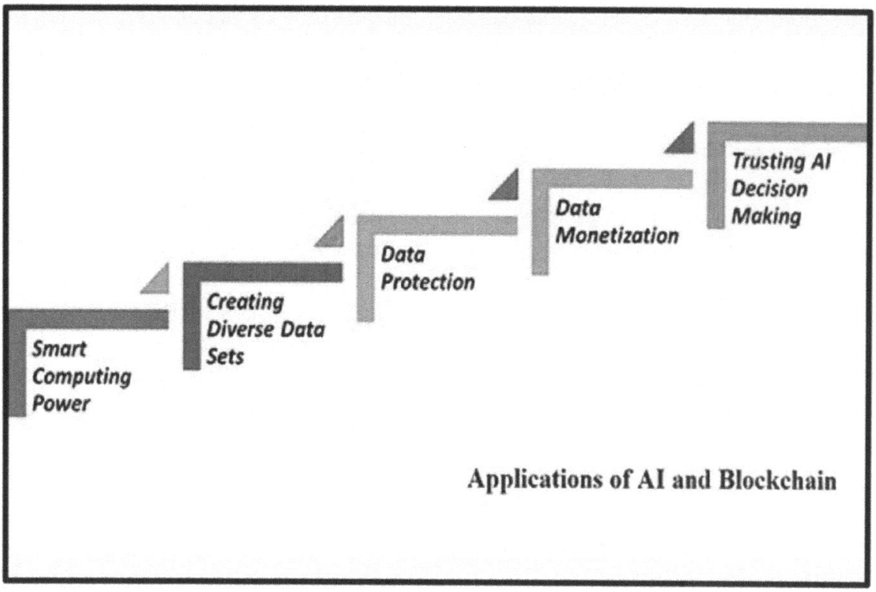

FIGURE 5.2 Applications of AI and blockchain.

5.1.4 Applications of Blockchain and AI

5.1.4.1 Smart Computing Power

On a server, running a blockchain and all of its cryptographic data will require a lot of computing power. For example, hashing algorithms are used to mine the Bitcoin blocks using the "brute force" technique that involves manually identifying all potential solution possibilities and testing if all satisfy all arguments of the problem until confirming the given transaction. We will step on from this with AI and approach projects in another way with increased intelligence in an effective manner. Consider an ML-based algorithm that, with the right training data, could virtually polish its skills in "real time."

5.1.4.2 Creating Diverse Datasets

Dissimilar to computerized reasoning-based projects, blockchain innovation makes decentralized, straightforward organizations that can be accessed or achieved by anybody, all throughout the planet in open blockchain networks circumstance. While blockchain innovation is the record that powers digital forms of money, blockchain networking is presently applied to the various ventures for carrying out decentralization. For instance, singuarlity NET is explicitly centered on utilizing blockchain innovation to empower a more extensive dispersion of information and calculations, guaranteeing the future improvement of man-made reasoning and the making of "decentralized A.I."

Singularity NET blends blockchain and AI to build smarter, open AI along with blockchain networks capable of hosting a wide range of datasets. The intercommunication of AI agents could be allowed by building an API of APIs on the blockchain. As a result, a broad variety of algorithms can be developed using a wide range of datasets (Ekramifard, A. et al. 2020).

5.1.4.3 Data Safeguarding

The development of AI is entirely based on the feedback of data which is our data. AI gathers knowledge from the environment and the details of what is going on in it. In essence, data is given to AI, and then AI has the ability to develop itself over time as a result of it. Blockchain, on the other hand, is a technology that facilitates the cryptographic storing of data on a spread-out ledger. It facilitates the development of completely encrypted databases which can only be accessed by some particular parties that are granted permission to do this. When blockchains and AI are combined, we have created a backup scheme for people's confidential and extremely important personal info. Medical or financial information is far too private to entrust to a sole company's algorithms. Recording all these data on the blockchain, which is accessible by AI with authorization and after it has been processed through the correct steps, will provide us with huge benefits such as customized reviews while securely storing our confidential data (Gill, S. S. et al. 2019).

5.1.4.4 Data Monetization

The monetization of data is another disruptive breakthrough that could be made feasible by merging the two technologies. For big corporations like Facebook and Google, monetizing gathered data is a major source of revenue. Data are being weaponized against humanity by allowing others to control how records are sold to generate money for companies. Blockchain enables us to encrypt our data and make it used in the manner that we see fit. It also helps us monetize data on a personal basis if we so wish, without our important valuable information being exposed. This is vital to comprehend in so that we can counteract skewed algorithms in the future and construct diverse datasets (Goertzel, B. et al. 2017).

The same can be said for AI programs that depend on our valuable records. For the mentioned AI algorithms which learn and change, AI networks are required to purchase data straight from developers from databases. This will change the whole process into a much more equitable process than it was as of now, with no tech behemoths betraying people. This type of data marketplace will also make AI more available to small businesses. Developing and providing AI are prohibitively expensive for companies that are not able to generate their own results. Via decentralized data servers, they acquire the ability to get data that is either too expensive or kept secretly (Hammi, M. T. et al. 2018).

5.1.4.5 Trusting AI Decision-Making

Since AI algorithms acquire intelligence through experience, data scientists would have a tougher time understanding how these systems arrived at simple conclusions. This is basically because AI algorithms have the ability to work with vast numbers of data and various variables. However, we should continue to scrutinize AI reports to ensure that they are reliable. They consist of permanent records of all the data, various variables, and all the executed processes used by AIs in their own decision-making processes thanks to blockchain technology. This makes auditing the whole process much simpler. The steps from the point of data entry to the point of conclusions can be monitored using blockchain programming, and the observers can be certain that the information has not been tampered with. This is an important step because people and companies can't start using AI apps until they understand how they basically function and understand the type of data they use to make decisions (Huh, S. et al. 2017).

5.2 BLOCKCHAIN, IMPROVING MACHINE LEARNING MODELS

Microsoft researchers are collaborating to build ML systems that are hosted on public blockchains. Since blockchain enables people to be compensated for helping to improve models, this relationship is rewarded. Despite the fact that ML is progressing at a rapid rate, the benefits are not yet widely available. People with limited resources cannot often be able to access cutting-edge ML systems, which are highly centralized and dependent on proprietary databases that are difficult to replicate. Furthermore, if models aren't retrained with new data on a regular basis, they become obsolete.

Microsoft is attempting to make AI decentralized and interactive by using blockchain technology. In the future, people will be able to use common computers and applications (such as tablets, browsers, and smartphones) to run sophisticated ML algorithms and collaborate on data and model creation. Microsoft is developing a decentralized and collaborative AI on blockchain platform that will enable the AI group to collaborate on training models and creating datasets on public blockchains. The ML frameworks, however, are free to use. Personal assistants and recommender programs are only a couple of the many choices available (e.g., what Netflix uses to recommend shows). Ethereum was used to construct proofs of concept (Idelberger, F. et al. 2016).

Since blockchain provides participants with trust and security, it makes sense to use it. You should be totally confident in the code you're working with. Instead of using proprietary cloud services, Microsoft's architecture uses smart contracts to codify product specifications. Models may be modified on the blockchain or used off-chain on the user's local machine at no cost to the customer. The model will always work as planned due to the immutability of blockchain and smart contracts. Unless the model is updated and checked, every user will see it as the "only true version."

In addition, blockchain introduces a reward mechanism that allows users to contribute data that makes models develop. We can reliably compute and log changes since we can check and monitor changes. Contributions that boost AI models should be rewarded (in tokens). According to Microsoft analysts, updating a Perceptron model on Ethereum costs $0.25. They expect to eliminate the need for this fee in the future. Users are compensated based on how well their feedback aided the model's improvement. Positive donations are compensated, while poor (malicious) contributions are penalized by the deduction of user funds. While Microsoft's framework isn't yet operational at scale, its vision could become the standard in the near future. Increasing the pace of AI adoption and effectiveness by allowing advanced AI models and massive datasets to be widely exchanged, upgraded, and educated (Karuppiah, M. et al. 2014).

5.2.1 Some Examples of Blockchain and AI-Integrated Softwares

- *AICoin*—AICoin is basically a money conceptual idea in which a token represents the benefits of using AI. The engineers developed AI models to find out how to identify and exchange designs that are not shown in the dozen or more so mostly liquid digital money marketplaces in this venture. The explanation for this, according to the designers, is that in accordance with the developing team, AICoin aims to allow financial backers of the token to build abundance using the power of AI and blockchain.
- *Botchain*—Botchain is basically a blockchain-based project that both speeds up and simplifies the world of deceptively AI bots and apps. The role includes capabilities for general enlistment, character approval, bot analysis, and consistency. Anyone who uses AI-controlled objects, as well as those who build them, will benefit from the Botchain's simple frameworks.
- *DeepBrain Chain*—DeepBrain Chain is a blockchain-based AI-processing stage that is decentralized, easy to use, and safe. It's essentially a decentralized neural network. Its aim is to build a decentralized distributed computing network that will help AI grow. According to the team of engineers, the company plans to move from their currently used NEP-5 chain to a local substrate token with localized management. DeepBrain Chain also has the ability to secure an information exchange base that increases the importance of data all the while ensuring the data protection by isolating data ownership originating out of data use.
- *Matrix AI*—The matrix to follow through on the blockchain promise; AI uses AI-applied technologies, for example, NLP. Auto-coding astute agreements, AI-controlled online security, flexible blockchain borders, complex designation organization and that's just the beginning of the project's highlights. Furthermore, an AI-powered safe virtual machine recognizes planned escape clauses and pernicious goals while maintaining vigor under intense emphasis assault with a generative ill-disposed organization (Karuppiah, M. et al. 2015).
- *Numerai*—Numerai is a cryptocurrency project that focuses on AI and has the main problem of data science at its heart. It is essentially a competition in which one must deal with the problem of forecasting the financial exchange. You must create a model while using the model Python along with R contents in the Numerai competition. A new round begins every Saturday at 18:00 UTC, and new tournament's data is delivered. Similarly, the accommodation deadline is on Monday at 14:30 UTC.
- *Singularity-NET*—Singularity-NET is basically a full-stack AI system that runs on a regionalized protocol/process. It is claimed to be the only regionalized stage that allows AI to participate and promote at scale. This

Singularity-NET network is basically the backbone that enables AI administrations to connect and execute. Singularity-NET allows everyone to navigate a global-based network of AI calculations, administrations, and experts

- *Namahe*—Namahe is a decentralized development network focused on blockchain software and technology. It integrates AI and blockchain technology to create a stable ecosystem in which companies can also save money on excessive feedback and increase the productivity of their supply chains. Namahe improves supply chain efficiency by allowing the AI layering to continuously screen the store network, detecting for anomalies/problems in the examples to inference extortion, delaying, and unexpected happenings and hailing the data for audit.
- *Peculium*—Peculium is a completely straightforward and decentralized reserve funds at the board stage fuelled by AI and AI. Based on AI, the task means to boost benefits and reserve funds. It intends to assist its clients with beating the dangers of the present speculations industry. Peculium additionally observes and gives the board of cryptographic money resources for clients. This is accomplished by Peculium's cutting edge monetary consultant, AIEVE, for example, computerized reasoning, ethics, values, and equilibrium (Karuppiah, M. et al. 2019).

5.2.2 SingularityNET

SingularityNET, a nonbenefit association settled in Amsterdam that dispatched in 2017, happens to be an open commercial center for AI calculations. CEO Ben Goertzel, who is additionally on the chair of the OpenCog Foundation and also the AGIS (Artificial General Intelligence Society), founded the firm. SingularityNET is a commercial center where engineers and AI sellers can exchange their equipment or applications for other AI administrations or digital currencies. The money of SingularityNET is known as "AGI tokens." Smart agreements based on the blockchain, as indicated by the association, permit exchanges between market members. As per the Whitepaper, SingularityNET claims that their commercial center works by providing purchasers and vendors with a scope of normal AI programming and equipment administration APIs that can be incorporated into brilliant agreement models. SingularityNET offers the models also (Kevin, A. C. et al. 2018).

This may, for instance, support organizations and individual engineers to buy and sell enormous-scope AI administrations like:

- Picture and video preparing organizations: Such as those that perceive people in chronicles or make text depictions for a particular picture.
- Language dealing with organizations: Text layout and consultation, language understanding, or text-based idea assessment.
- Admittance to curated datasets to plan AIs: Marketplace customers may share data, for instance, establishment data (for instance, online media data) to help train AIs that would then have the option to be used for assessment of other datasets.

- Organizations could moreover request to have a particular dataset taken apart from various individuals in the business community.

SingularityNET reports that their funded ICO brought more than $36 million up in the primary moment of its culmination in spite of the fact that the venture was just in its beginning phases with nothing to show regarding obvious strong execution. SigularityNET says that their in-house innovation staff is inspired by working together diligently with groups from Hanson Robotics (producers of Sophia Robot), Mozi Health (zeroed in on biomedical AI), and iCog Labs, along with the organization's CEO (Krittanawong, C. et al. 2020).

5.3 DEEPBRAIN CHAIN

DeepBrain Chain is fundamentally a Singapore-based nonprofit organization with around 35 workers. The organization professes to be building up a cooperative AI-processing framework utilizing AI and blockchain innovation. As indicated by assets from Whitepaper in the DeepBrain Chain, the association plans to essentially give a business community to AI equipment figuring administrations through an organization in which all gadget individuals are isolated into independent testing "hubs" in light of their registering limit. These DeepBrain Chain network hubs might be enormous hubs like mining pools, which are broadly used to mine the DeepBrain Chain token (Kumar, R. et al. 2021).

To prepare algorithms, medium-sized hubs use cloud computing stages such as Microsoft Azure or presumably home PCs. These large or medium hubs might be rented out to endeavors or associations that need computational force for AI ventures. A medium-sized hub will incorporate a home PC framework, cloud suppliers, and human clients. Bigger weighty processing apparatuses utilized for DBC coin mining will normally shape the organization's bigger figuring hubs. Excavators or people who access the plan by downloading the DBC program, which haphazardly doles out them to a particular hub relying upon their preparing limit, make up the entire organization (Luu, L. et al. 2016). Figure 5.3 shows DeepBrain Chain architecture.

After the preparation is finished, the miner is approached to give a contribution to the innovation that was utilized to prepare the client's AI. They would then repay the coach in tokens for their administrations. On the site, diggers can even transfer and lease their own datasets. For instance, AI sellers may transfer information and models fundamental for neural organization calculation to the DBC-decentralized capacity organization, and at that point send registering solicitations to exploit the organization's assets. The picture in Figure 2.1 from DBC portrays the organization's plan of action structure:

DBC likewise says that their site can be utilized by organizations and scientists to trade datasets, just as an exchange stage for fake neural organizations. It's important that there is no verification that the organization had accomplished perceptible results. Dongyan Wang, the head of AI at DeepBrain Chain, moved on from the University of Wisconsin–Milwaukee with a PhD in Electronics and Computer Science. He's worked with Samsung, CIisci, and NetAPP, among others. He has been the Global AI Leader of Midea Group, a Fortune Global 500 organization in

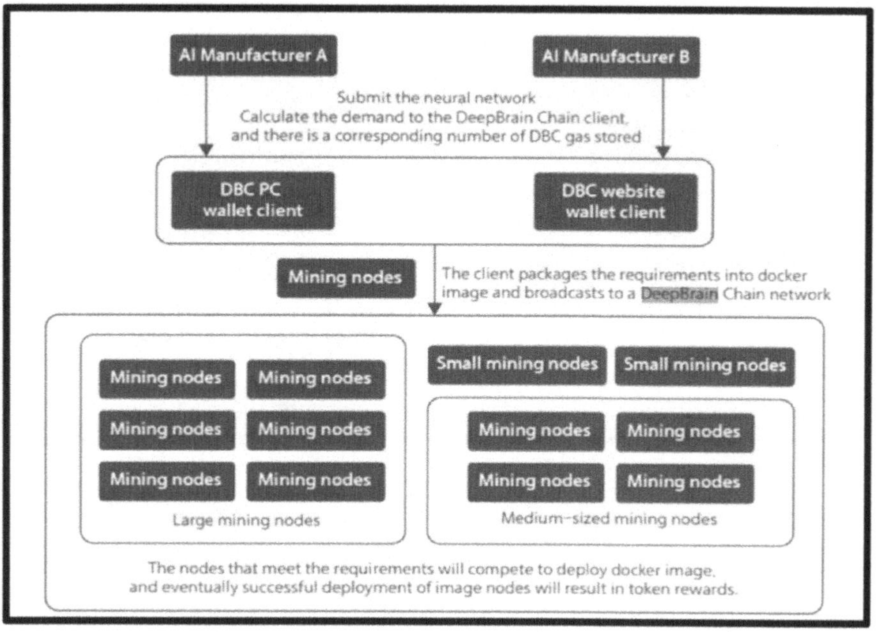

FIGURE 5.3 DeepBrain Chain architecture.

China. DeepBrain Chain has additionally banded together with SingularityNET to empower AI specialists on the Singularity NET commercial center to utilize the DBC commercial center's processing abilities. The association additionally is by all accounts selling computational "groups" to excavators everywhere in the world, professing to have acquired about $100 million in vows up until now; yet we couldn't freely approve this attestation (Lin, C. et al. 2019).

5.4 DISRUPTIVE INTEGRATION OF BLOCKCHAIN AND AI

On the one hand, blockchain has vulnerabilities including stability, scalability, and performance. On the other hand, AI has its own set of trustworthiness, explainability, and privacy concerns. The union of these two innovations is inescapable; they have the ability to revolutionize the next modern age by complementing each other. As seen in Figure 5.4, blockchain can provide AI with trustlessness, anonymity, and explainability; in turn, AI will assist in the development of a ML framework on blockchain for enhanced stability, scalability, personalization, and governance.

5.5 BLOCKCHAIN FOR AI

For various components of AI, such as data, algorithms, and computational resources, blockchain can power decentralized marketplaces and collaboration platforms. These would catalyze AI creativity and adoption at a never-before-seen scale.

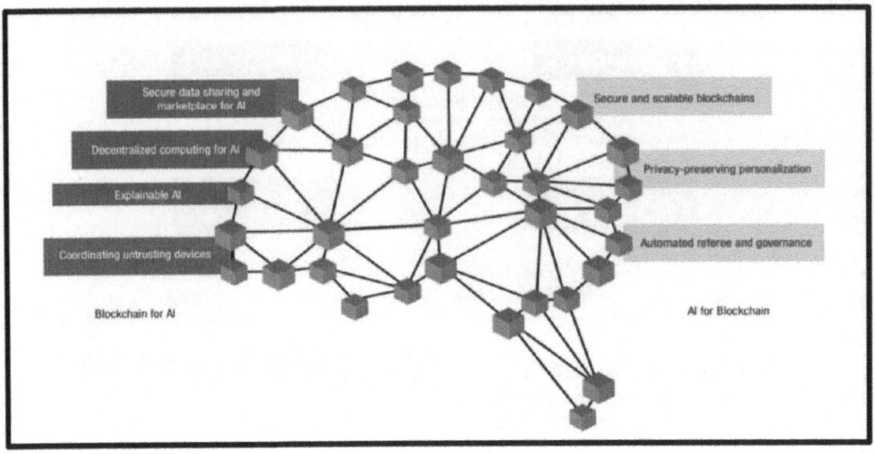

FIGURE 5.4 The integration of AI and blockchain: (a) blockchain for AI and (b) AI for blockchain.

AI decisions can now be more straightforward, explainable, and trustworthy, thanks to blockchain. Since all the records on the blockchain are public, AI is crucial for ensuring users' anonymity and privacy (Lin, C. et al. 2019).

5.5.1 Secure Data Sharing

The immense volume of information accessible for science, development, and trade is one of the main forces behind the new AI upset. In the present information-driven economy, information is the new gold. Be that as it may, there are significant difficulties in securing this gold. In any case, it could be a little difficult to get satisfactory data records for the off chance that you are working for companies like Facebook or maybe even Google. This in long run prevents the resistance by AI examiners and associations, which is required for helping AI. Second, assurance is fundamental and creates stress along with a movement pertaining to openings and maltreatment done on individual records. Unmistakably, the new Facebook shock, where 50 million customers profiled and centered not given any consent from the Cambridge Analytica, which is basically a political outcast firm/company. The zeroing in on direct is disturbing and stood out from "a redid murmur crusade: Groups both moral and malevolent can isolate Americans, murmuring into the ear of every single client, poking them dependent on their feelings of dread and urging them to murmur to other people who share those apprehensions" (Liu, J. et al. 2020).

5.5.2 Your Records/Data, Your Cost

Splitting old data and taking control of your own records, blockchain progressions can basically permit the user to sell their records through splendid arrangements.

The aforementioned information empowers information commercial centers without agents, making them safer and private. Such commercial centers will bring down the boundary for more modest players, leveling the battlegrounds and in this way encouraging developments. Through innovation, for example, zero-information evidences, organizations and analysts may look for significant data without knowing the subtleties of the information or the character of the information proprietors. We can't pressurize sufficiently for getting the colossal effect of having the option to sift through and find the information you need while keeping clients' security unblemished. An example, like Nebula Genomics, which is basically a start-up assisted with set up by Harvard University which has a part named George Church, which gives the business community, who interfaces people that need the genomes to be sequenced for associations that require those records. In a similar effort, the Longenesis gives a phase to divide and adjust life records like clinical evidence and prosperity records (Lopes, V. et al. 2018).

5.6 EXPLAINABLE AI

Regardless of the wide accomplishment of AI in building independent frameworks that are equipped for seeing, learning, and following up on their own, there is hesitance to embrace these frameworks. One explanation is that with AI procedures', for instance, profound understanding, it is hard to comprehend what precisely goes within the secret elements. Accordingly, choices made by those frameworks are unexplainable to human clients and along these lines can't be checked or believed. Wavering is fundamentally higher in the fields of clinical assessment and money-related organizing, where sensibility gets crucial as misguided decisions could result in life loss or monetary catastrophe. Basically, it's that we find a perpetual way that follows the headway of the record stream and complex acts of the AI-established structures. Blockchain has the ability to unequivocally do that, following each step in the dynamic chains and information handling. Through practices of AI-established structures on various different database and executed circumstances, the perception of and trust in the decisive steps made using those systems are attained. Far and away, superior human clients will have an unmistakable path to follow back the machine choice cycle, settling on defense of those choices a lot simpler. Moreover, it gives experience in tuning those secret elements to adjust execution and forecast precision with the reasonableness of the framework. In the event of disastrous occurrences, these blockchain-based paths will be fundamental to decide if people (and explicitly specifically) or man-made objects are to blame.

5.7 AI FOR BLOCKCHAIN

Thousands of parameters and trade-offs between security, efficiency, decentralization, and other factors go into the design and operation of a blockchain. AI can assist with these assessments as well as simplify and improve blockchain for improved efficiency and governance. Furthermore, since all records on the blockchain are transparent, AI executes a critical part in ensuring user anonymity along with safety.

5.7.1 Security and Scalability

But just in case that adversary guarantees the bigger part mining power, blockchain is for all intents and purposes hard to hack. The applications and functionalities dependent on top of the blockchain stage, deplorably, don't provide the same amount of security. An example, like the localized self-administering affiliation (D.A.O.), one major and greatest crowdfunding packs with around done 150 million dollars of computerized money called Ether, the main part of setback of a 50-million dollar burglary. The programmer misused a few slip-ups made in the composition of the keen agreements that permitted rehashed exchanges to be run that pulled out more cash than the asset put in. With the unimaginable advancement made by AI, a blockchain administered by a shrewd AI calculation could possibly identify the presence of assaults and naturally summon the proper protection instruments (Mamoshina, P. et al. 2018).

Exactly when the damage is unavoidable, the AI may on any occasion separate the attacked part from the blockchain stage, shielding the rest from the danger of an attack. The relative AI can be utilized to regulate the blockchain, making it extra flexible and amazing. For example, considering that there occurs a surge in the amount of trades, the AI could be adequately helpful to extend the square manufacturing increase growth, which can basically create the overall output to the detriment of longer affirmation setups

5.8 PRIVACY AND PERSONALIZATION

In the event that you stress over whether the any political decision will be undermined or whether your information is protected in interpersonal organizations, blockchain is for you. Blockchain can give you back authority over your own information. Yet, it accompanies an expense. In standard joined settings, as Facebook, YouTube, or streaming platform Netflix, assembled customer records/behavior are analyzed to modify the substance for customers. This is similar to Facebook where you log in to find the posts created by other users or by the colleagues you associate most with or enter into Netflix where you are given the choice of films that are more relatable considering your own taste. Taking back your security infers no company comprehends what you prefer—so basically, you will apparently have to go to various pages and platforms to find appropriate substance, without any help of modified customs.

However, is there another method to deal with to acquire both assurance and redid knowledge? Man-made insight comes to the rescue with another substance assurance model. A decentralized substance giver, like a relational association on blockchain, can utilize AI on behalf of the customers in order to tweak content (Mattos et al. 2020). AI software will basically run on customers' gadgets to look at their scrutinizing practices and relaxation exercises. Relatable substance that the customers will be attracted to, instead of pushing them, and appeared to all humans. Noting the fact that all the computation is basically executed in the same place—not up close and personal information anytime makes the customers' contraptions. Further, sanitization of customers' substance tendencies can be executed in order to hold content providers back from making profiles of customers. Thus, this new attraction—established model gives security as well as personalization immediately (Mauri et al. 2017).

Notwithstanding later and fast occasions, both AI and blockchain still have extensive, troublesome involvement with front of those people. For ML/AI, the latest improvement is the Google Duplex announcement, which basically can make customized mobile device choices for your well-being. It is exceptional until you read that the endeavor is limited to absolutely three things: bistro reservations, salon plans, and event hours. Likewise, paying little heed to AI's new, monstrous movements in computations and the colossal proportion of enrolling power and data, it can't remain mindful of the boundless difficulties of the living world. For blockchain, these new security events with Etherum, BitcoinGold, ZCash, and various different advanced types of cash suggest us that there is some time until we value both secured and flexible blockchains in genuine applications. Setting out toward the future, the companionship between blockchain and AI will give boundless progressions and prove helpful in the form of changes benefitting our overall population—perhaps they may endure until the very end (Mohanta et al. 2020).

5.9 CONVERGENCE OF BLOCKCHAIN AND AI WITH IOT

Blockchain advancement, the IoT, and ML/AI are currently seen as developments that can possibly change existing strategic approaches, make new game plans, and upset whole organizations. By giving a well-known and decentralized circled record, blockchain, for instance, will expand certainty, responsibility, security, and assurance of business measures. A blockchain or, all the more normally, an appropriated record may store an assortment of assets, like a log. Generally, these points of interest can be associated with cash and people. The IoT powers the mechanization of organizations and the convenience of business measurements, the two of which are basic (Montes et al. 2019).

Till this point, the linking between the aforementioned three progressions has been denied or not thought of, and blockchain, IOT, and ML/AI have been utilized independently. Be that as it may, these advances ought to be utilized pair, and they can, in the long run, meet up. One potential connection between these advancements is that IoT gathers and circulates information, blockchain gives the premise and builds up responsibility standards, and AI improves and controls cycles. These three headways are intended to be corresponding, and, when joined, they will abuse their full force. The mix of these advancements can be especially persuading for chiefs' information and the mechanization of business measures, which we dissect and address along with any remaining heads that can be lumped into a similar gathering.

5.10 CONVERGENCE OF BLOCKCHAIN, INTERNET OF THINGS, AND ARTIFICIAL INTELLIGENCE

Several years earlier, blockchain development was simply inspected concerning installments, i.e., with respect to Bitcoin and Ether. Fairly as of late, progressively more nonfinance-based use cases for the blockchain development has currently emerged, for instance, creation network of the chiefs and progressed characters. The later composing perceives the advantage of joining blockchain development with various headways like IoT and ML/AI. For example, it talks about the blockchain advancement usage to upgrade the structure establishment of various IoT gadgets (Nassar, M. et al. 2020). The format of the blockchains can be changed so much that the final output system is more ready to make and deliver IoT contraptions, mainly in regards to the quickness of trades. Other than maintaining focus on the blockchain tech in regards to IOT, a couple of assessments in like manner are base for the blend of blockchain tech along with ML/AI. Until this point on schedule, the accentuation is essentially on partner blockchain along with one or the other inventive development, as IOT and ML/AI, and also not having any critical bearing every one of the three headways meanwhile. Nevertheless, the certified ability of such new and also emerging advances may be proved if such improvements get added. The main plan is a blockchain-established structure that maintains IOT along with AI. Contrary to the aforementioned, this gives a non-specific diagram of the upsides of each progression and also how they supplement each other. The main and normal mix of blockchain advancement, IOT, along with AI appears with strong use case.

It's significant that these ideas that are introduced in many papers are relevant to public as well as private-made blockchains. The main fundamental contrast between the two sorts of blockchains is mainly that in a public-created blockchain, any part may get to the information put away on the blockchain. Passage to data in private blockchainsis limited to specific substances. Since the utilization cases might be completed on both public and private blockchain establishments, it is unimportant if the passageway is public or private for the reasons for this chapter. Moreover, it ought to be noticed that, similar to some other information base, blockchains are affected by defenseless data quality. This theme won't be facilitated in this chapter since it isn't carefully applicable to blockchain-based information management of data (Nugent, T. et al. 2016).

5.11 IMPROVING DATA STANDARDIZATION

Protection, security, and scalability of IoT gadgets, like keen home gadgets, splendid frameworks, sensors, robots, vehicles, or smart structures, gather a great deal of information. This information is regularly saved money on a fused laborer, where the information structure isn't standardized. Diverse legacy frameworks are utilized by various associations, making it hard to separate and translate information through all stages.

Blockchain development can help us with the similarity of records by making up a coordinated progressed stage for IoT records' accessibility for various get-togethers. Records will then be taken care of in one database plan. Since the application hash-limits usage, records on blockchain structures are conventionally taken care of in one data plan. In this way, data the chiefs could be upgraded by extended operation of set aside user records (Rodriguez-Mier, P. et al. 2016).

There exist basically two general amassing decisions for blockchain-established database, specifically on-server and off-server storage. On-server accumulating has the basic advantage that the main user data is reliably accessible on server and also can be taken back from available center point at whatever point. Regardless, accumulating requirements that are basic, which basically can incite "blockchain expanding," are a great deal of on-server-set aside database destroying blockchains' performance as well as flexibility. Off-server amassing gives an elective that basically stores the genuine database off server and simply keeps the main aggregated/summation of metadata on server. This alternate approach has the upside of essentially being more adaptable than on-server course of action; yet lessens data straightforwardness (Punithavathi, P. et al. 2019).

Another component of blockchain stages is the genuine degree of data insurance that can be done by the basic cryptography. On blockchains, trades are basically driven using nom de plumes are—in some blockchain structures like Monero or Zcash—drove absolutely anonymously. The plan of blockchain systems also considers full encryption of set aside and sent data so much that solitary the genuine contraption can scrutinize and create its own data through private/public key establishment. In IoT, machines and contraptions store a ton of sensitive data. It is crucial to ensure the assurance and security of this data. Today, IoT data is consistently sent directly from the machine to the specific informational collection (regularly a cloud-based platform), where the user records are recombined together and stored. Nevertheless, these user records aren't encoded and do, in like manner, not assurance security. Blockchain development can give immense benefits in such manner as blockchain advancement can without a very remarkable stretch assurance security of the accumulated data. Blockchain advancement has been made going with the procedure: security by a plan (Saleem, J. et al. 2018).

Furthermore, a blockchain is basically operationally adaptable and has an OK of hacks. This critical level of security rises out of the blend of cryptography and the understanding segments used. Thus, data security can grow using blockchain development. In any case, there is a trade-off between a huge level of assurance and control for illicit activities. If a blockchain stage is set up absolutely covertly, it is silly to hope to associate a trade with a particular social event. This anonymity opens the doorway for unlawful activities, for instance, tax avoidance or mental assailant financing. Reproduced insight can basically help extending security along with recognizing strange activities. A paper by Yin in 2019 proposed to utilize AI using user records' assessment to lessen the peril of unlawful activities that are taking place in blockchain coming about in light of the mystery of trades. Reenacted insight advancements benefit by the high proportion of gave IOT main records of data ever since AI computations acquire from the stored records—the more these records are used to set up the AI figuring, the more interesting and more thought out the presentation of the estimation is (Sgantzos, K. et al. 2019).

Today, the central drawbacks of IOT are its inability to store along with massages a ton of user records. The organization of data can be converted to be more versatile by a gathering of progressions utilizing blockchain development along with AI. Foes of these blockchain advancement reproach that these blockchain structures need flexibility because of the usage of all these energy-consumption understanding segments to support trades, for example, proof of the work arrangement. Regardless, there basically are distinctive more than normal energy-capable understanding frameworks, similar to affirmation of-stake or check or-authority which basically can assemble flexibility. In all honesty, all such high-energy usage will after a short time be a trinket of the Bitcoin association. Nearby blockchain, the advancement in AI can maintain a much more extensive extension in flexibility. Liu proposed a presentation in his work, smoothing out structures for such integrated blockchain-engaged IOT software/hardware. This structure could possibly be based on significant help learning, one kind of AI, to show up at a more critical degree of overall performance. The makers suggested us a "DRL-based estimation to intensely pick/change the square producers, arrangement figuring, block size, and square range to improve the presentation." To conclude this topic, the blockchain advancement can upgrade user record usage by the leading group of IOT gadgets because of its straightforwardness, trustworthiness, changelessness, secure structure, and insurance characteristics. Together with AI, it basically can address all the current limitations present with the IoT database (Shafagh, H. et al. 2017).

5.12 AUTHENTICATION IN ACCORDANCE TO A BLOCKCHAIN-ESTABLISHED IDENTITY

Additionally, blockchain advancement can be executed to confirm IT networking individuals and also can construct trust among each other by basically managing the character of IoT gadgets. Noting that character the board can insinuate individuals and associations notwithstanding—with respect to IoT—furthermore to IoT contraptions and human-made objects. Blockchain-established characters make sure that trade groups get a modernized character for blockchain, considering genuine character of those users (e.g., character card for various individuals and business registration segment of associations). Considering this type of a character, trades between an individual and an association (model: vehicle sharing) yet moreover between an individual user and a man-made machine (model: voyager transportation of a free vehicle) or between two man-made machines (model: self-administering vehicle paying for leaving) could be arranged beneficially—that is, with very quick trade speed and lower trade costs (Singh, S. et al. 2020).

Later on, the cash will eventually be moved between end users, associations, gadgets, and man-made machines. As shown by checks by IoT network analysis, greater than 20 billion contraptions are going to be related to the web by the year 2025 (IoT Analysis team). These particular devices will midway, in like manner, participate in portion measures. Along these lines, an absolutely new, conceivably decentralized, portion establishment wll be required. Individuals, associations, and man-made machines ought to be enlisted with basically their modernized characters on various blockchain structures. Accordingly, character heads of particular blockchain software will expect a vital key part (Swan, M. 2015).

Clearly, all such characters should be given and directed in conformity with user data laws for protecting their data. Regardless, the notion that these blockchains can't enough address data protection by arrangement isn't reasonable. This is a result of the way that the rapid blockchain advancement, with the consolidated induction systems along with the encryption measures, is shockingly upgraded version compared to nonblockchain-established structures prepared to first thing secure data by design, other than setting up the obligation regarding and, third engage the transformation of data. Another vital advantage of utilizing blockchain advancement is that the invariable management of the mechanized character is hard to create. With respect to self-sufficiently interfacing machines and contraptions, it is fundamental to have the alternative to rely upon the character of various things which can be realized by basically utilizing the help of blockchain advancement (Chen, T. et al. 2018).

5.13 AUTOMATIZATION BY MEANS OF SMART CONTRACTS

Close to database and character heads motives, the get-together of three headways by the application of blockchain along with IoT and also integrating AI commonly should be amazingly encouraging for the automatic processing of all the business measures. One of the huge pieces of interfacing these three progressions is the utilization of sharp arrangements. Sharp arrangements decide a lot of assurances, cautiously, in a show that normally executes the states of the understanding. In coding words, splendid arrangements are equivalent to "accepting by then" limits that describe unequivocal exercises if a unique task occurs. For example, one can consider a condition if for the circumstance the fair movement has been productive (if), a portion is carried out subsequently. With everything taken into account, splendid arrangements are the principal combination of the three design stages of IoT along with AI and also integration of blockchain advancement (Dinh, T. N. et al. 2018).

In spite of its gigantic prospective, sagacious agreements are as of now not utilized with respect to mechanical associations. The major issue behind conventional sharp arrangements is that they basically need crypto assets like Ether or even EOS and, along these lines, move proportions of such crypto assets. Regardless, associations are hesitant to utilize these aforementioned crypto assets, generally on account of regulatory and financial reasons. One major requirement is the unreasonable expense shakiness of crypto assets. Expecting a savvy arrangement is driven in Ether, where the tolerant party is introduced to a high change scale risk. Now and again, the expense of crypto assets augments or lessens by abundance margin of 10% within a day. Irrespective of the fact whether or not stable coins can offer a response for the high unconventionality of "conventional" cryptocurrency assets, they usually are enthusiastically used by mechanical associations or in back to back settings for the following reasons: First, the particular stable coins are as of now not regulated. Thus, danger went against associations don't attempt to utilize some of these on-regulated devices. Additionally, financing and information technology structures of associations are assigned to and working along with fiat money-related principles like the European currency Euro. Subsequently, it is processing load for associations to change over the stable coin usage in the "system-based" money. This change uses the time of both workforce and accounting resources (that is, trade costs, supporting for esteem differences). Figure 5.5 shows smart contracts overview.

The lone method/ability of these smart arrangements can be totally manhandled by utilizing blockchain-established fiat money that "travels through" the wise understanding. Simply a blockchain-established progressed European currency would enable Euro-named astute arrangements, with the ultimate objective that these

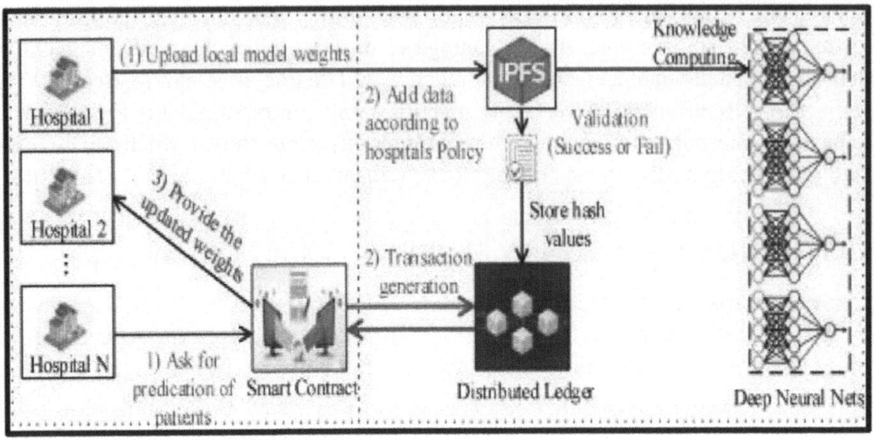

FIGURE 5.5 Smart contracts overview.

man-made machines, vehicles, or sensors can offer organizations their remuneration along with leasing and consideration. In view of a modernized blockchain-based Euro, such new strategies could show up: totally atomized devices making decisions all alone using AI and "fiscally making due" all alone using blockchain for money-related trades while executing an advantage place reasoning on the end instrument level (Dinh, T. T. A. et al. 2018).

The advantages of using such a DLT-established automated European currency are unpredictable. In any case, with this blockchain-established modernized cash, small payments for these IoT instruments can be executed with lower trade charges which are significant for the additional advancement of IoT. Additionally, all the trade named in this blockchain-established electronic Euro is going to be associated with internal ERP structures and will, henceforth, be open for financing and recording purposes. Third, interestingly with cryptocurrency assets along with stable coins, fiduciary money-related guidelines will get dreary, saving critical resources in this way. Fourth, especially automated fiat cash would adjust to current rules. Figure 5.6 shows smart contract testing using AI knowledge. There are at first new organizations that have made blockchain-established fiduciary currency-related structures and use electronic currency licenses for the vital tokenization of fiduciary financial structures. In this way, current associations mentioning all these blockchain-established Euro plans don't have to be afraid of authoritative weakness since current electronic money frameworks are utilized.

The blockchain-made European currency can be given either directly from the banks, electronic money associations, as shown in Figure 5.2, nonregulated foundations, or public banking institutions. As demonstrated by an assessment by BIS, more than 70 public banks in general as of now take apart the issued announcement of self-owned public bank-mechanized money (CBDC). No public bank has as of now introduced this kind of cash flow, irrespective of the fact that whether or not the Swedish public bank along with the Chinese public bank are initiating and may possibly dispatch a first public bank-mechanized money soon (Tanwar, S. et al. 2019).

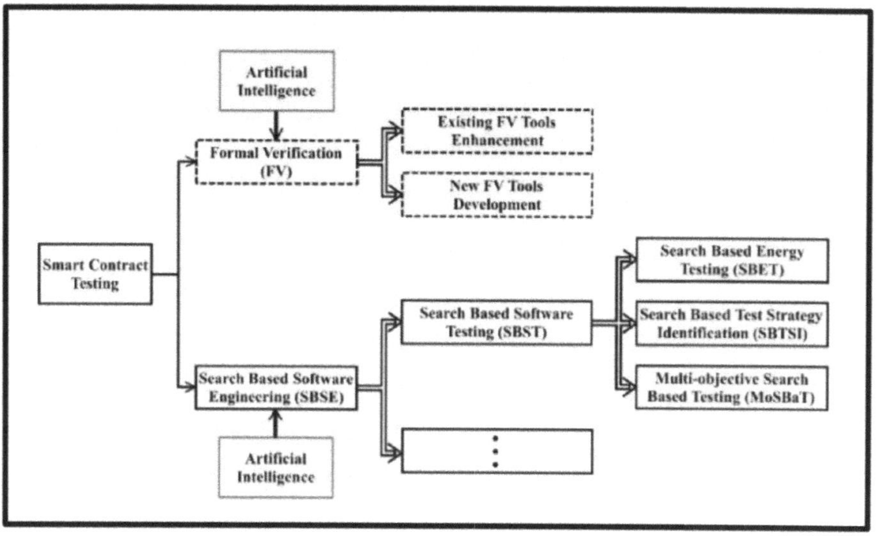

FIGURE 5.6 Smart contract testing using Artificial Intelligence knowledge.

The Chinese CBDC project DC/EP is currently in execution of a variety of Chinese metropolitan networks, for Chinese helpers of worldwide associations, similar to McDonald's or Starbucks. Until this point on schedule, the ECB has not yet pronounced the issuance of a Chinese Database Project. Eventually, a European blockchain-established Chinese-based project would be needed for the usage of a public finance house-moved European currency for insightful arrangements in the European business.

One may ask the following question: Why is a public finance institution-given Euro fundamental if electronic money givers have adequately initiated a blockchain-established Euro? The suitable reaction lies in the given details: This money given by electronic money suppliers considers electronic money/business finance house money, while the cash given by the public financing institution is public bank cash. Irrespective of the fact whether the two kinds of money address the Euro or not, by virtue of complete bankruptcy, business financing institution money could be reinstated, while public bank money is somethingthat basically can, by financial definition, not come up short. Whether or not this differentiation gives off an impression of being trifling amidst financial and money-related strength, it gets especially material amidst crisis.

5.14 INTEGRATION OF BLOCKCHAIN AND AI FOR MEDICAL SCIENCES

5.14.1 AI for Heart Medicine

AI is a quickly growing computer-processing request that can orchestrate very non-understandable data to make exact assumptions. PC-based insight has had striking victories in voice along with facial and picture affirmation in game-playing, in a variety of mechanical and intelligent fields, and is as of now being executed into clinical benefits. In heart medicine, AI basically can overview the function of the heart through imaging, understand heart temperament and limit from the ECG, and take some vital medical decisions similar to experts. In any case, an unseen assurance of AI is to control redone cardiovascular courses of action by describing novel totals past traditional ailment issues, improving outcome estimates, and individualizing treatment. In spite of the way that AI seems, by all accounts, to be prepared to comprehend this vision of precision medicine, particularly with the climb of wearable sensors and omic propels, progress has been mixed with data. A huge bottleneck is the shortage of colossal, secure, heterogeneous, and granular enlightening assortments, with exact improvements in proof, in people in peril. This obstruction is an unquestionably seen hindrance for AI in cardiovascular medication (Chen, W. et al. 2019).

5.14.2 Blockchain in Cardiovascular Medicine

Theoretical/hypothetical clinical benefits of a blockchain that basically initiates data that are as of now assembled and stored, are utilized by autonomous accomplices, and are consistently difficult to be accessible by the principal data provider (the patient). Blockchain is basically a data-driven model that usually tracks the ownership/user's activities as user records are executed between accomplices. Center points of the blockchain network check data moving between accomplices by understanding to make a cryptographic extraordinary imprint (hash) of data trades (Chen, W. et al. 2018). Figure 5.7 shows AI integrated with blockchain for medical purposes.

5.15 CURRENT APPLICATIONS OF INTEGRATED BLOCKCHAIN AND AI

There are a couple of executions that consolidate blockchain and AI. The American Heart Association has joined forces with the OHN to create AI integrated with

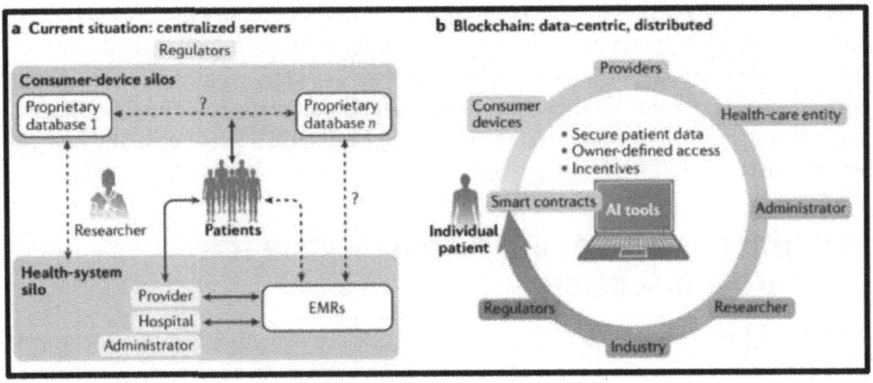

FIGURE 5.7 AI integrated with blockchain for medical purposes.

blockchain items like PatientSphere, a blockchain-established, HIPAA-agreeable information-sharing stage that utilizes AI to give care plans and exercise tips. As a rule, fire-up adventures work in three particular zones. To start, they use blockchain technology to capture and record information for the production of an AI architecture that investigates economics, heart imaging, and different contributions to foresee occasions, for example, outrageous myocardial-limited necrosis. Second, AI estimations and information ought to be decentralized. For instance, cooperation between the ObeN and MesStar Health Institute screens ill people with cardiovascular illness and gives motivating forces to the individuals who take an interest in wellbeing exercises. Third, to utilize blockchain as the spine for an AI-associated organization of sensors to anticipate cardiovascular illness, like Farasha Labs' exercises or the Health2Sync diabetes data trust.

5.16 THE PROSPECTIVE OF BLOCKCHAIN, IOT, AND AI IN COMBINATION

Considering the latest discussions, it's obvious that the mix of the rapidly advancing blockchain development with IoT instruments along with AI can open new game plans for the transformation of IoT contraptions. One aforementioned use case is briefed in the following.

One can consider a light (like a streetlight) that consists of its own blockchain-established character and works with a blockchain-established Euro. Along these lines, the light captures the surroundings with a self-administering substance working "in isolation." By using savvy arrangements, small payments can be done for the maintenance of the light, setting off the light to turn on. The light shimmers once some user/colleague pays for the light, such as an individual, an association, the strategy executioner, or someone else coming under this category. In this particular circumstance, pay-per-use portion plans could be done. Since the light has a high-level wallet, it can go probably as its own advantage local area (Wu, F. et al. 2018).

Given that all lights are related to a particular blockchain, they would be able to store user records, for example, about their use, execution, along individual time. Man-made awareness could utilize these user records and overhaul the association's help. For example, it could propose a much more standard upkeep of lights that are utilized routinely similarly as fast transmit the care group if there ought to be an event of a weakness. Moreover, AI has the ability that it can smoothen up the upkeep cycle by upgrading the mentioning cooperation of new tasks for associations or by simply helping with expecting the amount of new fragments needed even more accurately. This assistance would finally achieve less than home season of the association (Zheng, Z. et al. 2019).

Since lights can be tokenized as strengths, they have the ability to be made open to monetary supporters. Hence, monetary sponsor can develop and keep up these lights on a large scale. Thus, monetary sponsor will basically get the proposal on the lights' advantages. This application has the perspective of being a gamechanger. The tokenization of such strengths of the software has the ability to drive another surge of theories since monetary sponsor would be repaid directly with a segment of the appearance of the splitted asset, for the present circumstance, of the light. The convenience of tokenization doesn't simply uphold for lights anyway, but for all IOT [55–60] contraptions and, consequently, a wide extent of currently applied areas. For example, same can be said of sensors, vehicles, and man-made machines, various cameras along with trucks once these instruments are related with the web and with a blockchain-established network.

5.17 CONCLUSION

Notwithstanding quicker turn of events, combining both AI and blockchain has a lengthy, difficult experience of advancement. With AI, the high-level latest improvement is Google Duplex, which includes robotizing calls and making required errands for the clients. In any case, there is a constraint that can be utilized uniquely for accomplishing three principal capacities: occasion hours, café reservation, and boutique arrangements. Indeed, there are numerous headways including calculations that utilize more noteworthy measures of machine-figuring force and preparing information, while staying aware of the intricacies of this present reality is very troublesome.

While with blockchain innovation, subsequent to noticing the security-related penetrates including BitcoinGold, Etheru, ZCash, and a lot more digital forms of money grandstand, there will be a little postponement to create both secure and versatile blockchains utilizing certifiable applications. Organizations would be able to make unlimited advances and transitions as a result of their possession of blockchain and AI in the future. A significant portion of this mechanical seeing necessitates advanced learning stages such as upGrad and courses like Master of Science in ML and AI, which covers 20 programming languages, software, and libraries to provide active programming knowledge.

REFERENCES

Angraal, Suveen, Harlan M. Krumholz, and Wade L. Schulz. "Blockchain technology: Applications in health care." *Circulation: Cardiovascular Quality and Outcomes*, 10(9) (2017).

Bagga, P., A.K. Sutrala, A.K. Das, and P. Vijayakumar. "Blockchain-based batch authentication protocol for internet of vehicles." *Journal of Systems Architecture*, 113(101877) (2021).

Benchoufi, M., R. Porcher, and P. Ravaud. "Blockchain protocols in clinical trials: Transparency and traceability of consent." *F1000Research*, 6(66) (2017).

Bera, B., A.K. Das, M. Obaidat, P. Vijayakumar, K.F. Hsiao, and Y. Park. "AI-enabled blockchain-based access control for malicious attacks detection and mitigation in IoE." *IEEE Consumer Electronics Magazine* (2020).

Chen, T. et al. "Understanding Ethereum via graph analysis." In *IEEE INFOCOM 2018—IEEE Conference on Computer Communications* (2018), pp. 1484–1492.

Chen, W. et al. "Market manipulation of Bitcoin: Evidence from mining the Mt. Gox transaction network." In *IEEE Conference on Computer Communications* (2019), pp. 964–972.

Chen, W., Z. Zheng, J. Cui, E. Ngai, P. Zheng, and Y. Zhou. "Detecting ponzi schemes on Ethereum: Towards healthier blockchain technology." In *Proceedings of the 2018 World Wide Web Conference (WWW '18)* (2018), pp. 1409–1418.

Chen, Xuhui, Jinlong Ji, Changqing Luo, Weixian Liao, and Pan Li. "When machine learning meets blockchain: A decentralized, privacy-preserving and secure design." In *2018 IEEE International Conference on Big Data (Big Data)*. IEEE (2018), pp. 1178–1187.

Cong, Lin William, and Zhiguo He. "Blockchain disruption and smart contracts." *The Review of Financial Studies* 32(5) (2019), pp. 1754–1797.

Croman, K., C. Decker, I. Eyal, A.E. Gencer, A. Juels, A. Kosba, A. Miller, P. Saxena, E. Shi, E. Gün Sirer, D. Song, and R. Wattenhofer. "On scaling decentralized blockchains 9604." (2016), pp. 106–125.

Dinh, T. T. A., R. Liu, M. Zhang, G. Chen, B. C. Ooi, and J. Wang. "Untangling blockchain: A data processing view of blockchain systems." In *IEEE Transactions on Knowledge and Data Engineering*, 30(7) (2018), pp. 1366–1385.

Dinh, T. N. A., and M. T. Thai. "AI and blockchain: A disruptive integration." *Computer*, 51(9) (September 2018), pp. 48–53.

Ekramifard, Ala, Haleh Amintoosi, Amin Hosseini Seno, Ali Dehghantanha, and Reza M. Parizi. "A systematic literature review of integration of blockchain and artificial intelligence." *Blockchain Cybersecurity, Trust and Privacy* (2020), pp. 147–160.

Gill, Sukhpal Singh, Shreshth Tuli, Minxian Xu, Inderpreet Singh, Karan Vijay Singh, Dominic Lindsay, Shikhar Tuli et al. "Transformative effects of IoT, blockchain and artificial intelligence on cloud computing: Evolution, vision, trends and open challenges." *Internet of Things* 8 (2019).

Goertzel, Ben, Simone Giacomelli, David Hanson, Cassio Pennachin, and Marco Argentieri. "SingularityNET: A decentralized, open market and inter-network for AIs." *Thoughts, Theories Study Artificial Intelligence Research* (2017).

Hammi, Mohamed Tahar, BadisHammi, Patrick Bellot, and Ahmed Serhrouchni. "Bubbles of trust: A decentralized blockchain-based authentication system for IoT." *Computers & Security* 78 (2018), pp. 126–142.

Huh, S., S. Cho, and S. Kim. "Managing IoT devices using blockchain platform." In *2017 19th International Conference on Advanced Communication Technology (ICACT)*. IEEE (2017), pp. 464–467.

Idelberger, Florian, Guido Governatori, RégisRiveret, and Giovanni Sartor. "Evaluation of logic-based smart contracts for blockchain systems." In *International Symposium on Rules and Rule Markup Languages for the Semantic Web*. Cham: Springer (2016), pp. 167–183.

Karuppiah, M., A.K. Das, X. Li, S. Kumari, F. Wu, S.A. Chaudhry, and R. Niranchana. "Secure remote user mutual authentication scheme with key agreement for cloud environment." *Mobile Networks and Applications*, 24(3) (2019), pp. 1046–1062.

Karuppiah, M., and R. Saravanan. "A secure authentication scheme with user anonymity for roaming service in global mobility networks." Wireless Personal Communications, 84(3) (2015), pp. 2055–2078.

Karuppiah, M., and R. Saravanan. "A secure remote user mutual authentication scheme using smart cards." *Journal of Information Security and Applications* 19(4–5) (2014), pp. 282–294.

Kevin, A.C., E.A. Breeden, C. Davidson, and T.K. Mackey. "Leveraging blockchain technology to enhance supply chain management in healthcare: An exploration of challenges and opportunities in the health supply chain." *Blockchain Healthcare Today* (2018).

Krittanawong, Chayakrit, Albert J. Rogers, Mehmet Aydar, Edward Choi, Kipp W. Johnson, Zhen Wang, and Sanjiv M. Narayan. "Integrating blockchain technology with artificial intelligence for cardiovascular medicine." *Nature Reviews Cardiology* 17(1) (2020), pp. 1–3.

Kumar, Rajesh, WenYong Wang, Jay Kumar, Ting Yang, Abdullah Khan, Wazir Ali, and Ikram Ali. "An integration of blockchain and AI for secure data sharing and detection of

CT images for the hospitals." *Computerized Medical Imaging and Graphics* 87 (2021), p. 101812.

Lin, C., D. He, N. Kumar, X. Huang, P. Vijayakumar, and K.K.R. Choo. "Homechain: A blockchain-based secure mutual authentication system for smart homes." *IEEE Internet of Things Journal*, 7(2) (2019), pp. 818–829.

Liu, J., X. Li, Q. Jiang, M.S. Obaidat, and P. Vijayakumar. "Bua: A blockchain-based unlinkable authentication in vanets." In *ICC 2020–2020 IEEE International Conference on Communications (ICC)*. IEEE (2020 June), pp. 1–6.

Lopes, Vasco, and Luís A. Alexandre. "An overview of blockchain integration with robotics and artificial intelligence." *arXiv preprint* (2018).

Luu, L., D.-H. Chu, H. Olickel, P. Saxena, and A. Hobor. "Making smart contracts smarter." In *Proceedings of the 2016 ACM SIGSAC Conference on Computer and Communications Security (CCS '16)* (2016).

Mamoshina, Polina, Lucy Ojomoko, Yury Yanovich, Alex Ostrovski, Alex Botezatu, Pavel Prikhodko, Eugene Izumchenko et al. "Converging blockchain and next-generation artificial intelligence technologies to decentralize and accelerate biomedical research and healthcare." *Oncotarget* 9(5) (2018), p. 5665.

Menezes Ferrazani, Francine Krief, and Sandra Julieta Rueda. "Blockchain and artificial intelligence for network security." (2020), pp. 101–102.

Mohanta, Bhabendu Kumar, Debasish Jena, UtkalikaSatapathy, and Srikanta Patnaik. "Survey on IoT security: challenges and solution using machine learning, artificial intelligence and blockchain technology." *Internet of Things* (2020).

Montes, Gabriel Axel, and Ben Goertzel. "Distributed, decentralized, and democratized artificial intelligence." *Technological Forecasting and Social Change* 141 (2019), pp. 354–358.

Nassar, M., K. Salah, M.H. UR Rehman, and D. Svetinovic. "Blockchain for explainable and trustworthy artificial intelligence." *WIREs Data Mining Knowledge Discovery* (2020).

Nugent, T., D. Upton, and M. Cimpoesu. "Improving data transparency in clinical trials using blockchain smart contracts." *F1000Research*, 5 (2016), p. 2541.

Punithavathi, P., S. Geetha, M. Karuppiah, S.H. Islam, M.M. Hassan, and K.K.R. Choo. "A lightweight machine learning-based authentication framework for smart IoT devices." *Information Sciences*, 484 (2019), pp. 255–268.

Rodriguez-Mier, P., C. Pedrinaci, M. Lama, and M. Mucientes. "An Integrated Semantic Web Service Discovery and Composition Framework," *IEEE Transactions on Services Computing*, 9(4) (2016), pp. 537–550.

Saleem, Jibran, Mohammad Hammoudeh, Umar Raza, Bamidele Adebisi, and Ruth Ande. "IoT standardisation: Challenges, perspectives and solution." In *Proceedings of the 2nd International Conference on Future Networks and Distributed Systems* (2018), pp. 1–9.

Sgantzos, Konstantinos, and Ian Grigg. "Artificial intelligence implementations on the blockchain. Use cases and future applications." *Future Internet* 11(8) (2019), p. 170.

Shafagh, H., et al. "Towards blockchain-based auditable storage and sharing of IoT data." In *Proceedings of the 2017 on Cloud Computing Security Workshop, CCSW 2017.* ACM (2017), pp. 45–50.

Singh, Saurabh, Pradip Kumar Sharma, Byungun Yoon, Mohammad Shojafar, Gi Hwan Cho, and In-Ho Ra. "Convergence of blockchain and artificial intelligence in IoT network for the sustainable smart city." *Sustainable Cities and Society*, 63 (2020), p. 102364.

Swan, M. "Blockchain thinking: The brain as a decentralized autonomous corporation." *IEEE Technology Society Magazine*, 34 (2015), pp. 41–52.

Tanwar, Sudeep, Qasim Bhatia, Pruthvi Patel, Aparna Kumari, Pradeep Kumar Singh, and Wei-Chiang Hong. "Machine learning adoption in blockchain-based smart applications: The challenges, and a way forward." *IEEE Access* 8 (2019), pp. 474–488.

Wu, F., L. Xu, X. Li, S. Kumari, M. Karuppiah, and M.S. Obaidat. "A lightweight and provably secure key agreement system for a smart grid with elliptic curve cryptography." *IEEE Systems Journal*, 13(3) (2018), pp. 2830–2838.

Yaga, D., Mell, P., N. Roby, and K. Scarfone. "Blockchain technology overview, retrieved from National Institute of Standards and Technology (NIST)." U.S. Department of Commerce (2018).

Zheng, Zibin, Hong-Ning Dai, and Jiajing Wu. "Blockchain intelligence: When blockchain meets artificial intelligence." *arXiv preprint* (2019).

6 Big Data Analytics and Machine Learning

Francis Alex Kuzhippallil, Adith Kumar Menon, Ramani S, and Marimuthu Karuppiah

CONTENTS

DOI: 10.1201/9781003152392-6

6.1 INTRODUCTION: BACKGROUND AND DRIVING FORCES

Big data has revolutionized the decision-making process of various business organizations. It has helped uncover several information that would have been hidden otherwise. Various techniques and tools used in big data support in providing meaningful insights about various patterns and trends for data of any size, structure, and source. The advent of cloud computing has increased the computing power and automation ability. This has led to much efficient operations in a wide range of real-time applications from fraud detection to customer personalization. The fusion of data visualization, machine learning (ML) models, and big data has made it possible to answer more advanced business intelligence queries instantly as compared to traditional methods. Significant rise in data has only increased the growth of big data, and this torrential flood of data ensures that this field will never face extinction. Hence, knowing about this golden field is of utmost importance (Fong et al. 2018).

In 2005, Roger Mougalas coined the term big data. Even before then, the concept was applied in business which had then used spreadsheets to trend matching and number analysis. The biggest haphazard traditional data warehousing faced was the lack of parallelization of tasks due to which speed was compromised. In 2006, Hadoop framework was developed by Yahoo and launched as an Apache open source project. Their biggest feature was the distributed processing framework which enabled running big data applications on clustered platforms. During this time only, technological giants Google and Facebook managed to take advantage of big data analysis. By the 2010s, industries of various domains that include science, finance, healthcare, social networking all began to grasp the importance of being called a big data analytics company. Recently, Amazon Web Services (AWS) and various other cloud computing platforms made it much easier to use big data analysis opportunities (Manogaran et al. 2018).

6.2 SCOPE OF BIG DATA ANALYTICS

Big data analytics covers various business domains mainly healthcare, stocks, tourism, and much more. The scope of big data analytics is not limited to current technologies. Big data will exponentially increase as technological advancement occurs. Big data analytics has benefited many industries namely airline industry, banking, science and government, and healthcare.

6.2.1 Airline Industry

Airline industry revolves around a large volume of data such as traffic control, aircraft maintenance information, customer flight preference, tracking of flight, baggage handling, and so on. Since each of these data is real time and keeps changing daily, it is very important to have a quick analysis of the status, thereby making big data analytics one of their major focus.

6.2.2 Banking

Better utilization of unstructured data has been the biggest challenge which banks had faced over the past. Along with big data, banking has got an additional boost of better decision-making, the most important factor of a bank. Also, it has helped to reduce the fraudulence happenings as well (Tsai et al. 2015).

6.2.3 Science and Government

Big data analytics has been able to provide meaningful insights to various researchers which have helped them produce better results at a faster rate with less number of trials. Law enforcement obedience and public safety have increased with the introduction of big data analytics. It has also helped streamlining all their operations and produce better results at a faster rate.

6.2.4 Healthcare

With big data analytics, customer feedback can be easily visualized, and hence, it allows all the healthcare industries to improve their services and their organizational efficiency.

6.3 BIG DATA ANALYTICS TOOLS

Most NoSQL databases are used for big data analysis because of the enhanced capability for dynamic organization of unstructured data. Following tools are the most prominent ones used in the industries (Kambatla et al. 2014).

6.3.1 Apache Kafka

A scalable system that enables real-time publishing and consumption of a large number of messages upon subscription. It is also an open source which is used for data storing and streaming data analysis. Apache Kafka is log based, and it allows publishing data in any real-time applications.

6.3.2 HBase

This is data storage that is column oriented and runs on Hadoop Distributed File System. It's an open-source tool and is highly scalable. It shares similarity with the big table designed by Google, but Hbase has a cutting edge with high fault tolerance. It is optimal when random data is procured.

6.3.3 Hive

It's an open-source tool that runs in Hadoop file system. Hive is mainly used for structured data. Similarly, Hive is swift and highly scalable. To improve the performance, Hive uses the data partitioning technique. Hive is based on HQL.

6.3.4 Map Reduce

This is a software tool used for processing huge amounts of data (includes both structured and unstructured) in distributed computing. It is a highly scalable system. In addition, it possesses flexibility, parallel execution, and cost-effectiveness as other prominent features.

6.3.5 Pig

It is a large-scale open-source application on Hadoop clusters and used for parallel programming of Map Reduce. It is highly extensible and has a unique ability to handle heterogeneous data. Further, Pig contains a large repository of operators, thereby making programming effortless.

6.3.6 Spark

This is the distributed computing framework and open-source tool used for data analytics across cloud computing platforms. Reusability is one of the prominent features of Spark. Similarly, multilingual support is offered as well.

6.3.7 YARN

This mainly focuses on cluster management and is used in second-generation Hadoop. It works as a job scheduler and further allocates resources to each of them. It also functions as node resource controller that monitors each node and manages them.

6.3.8 Presto

It is a Facebook-powered SQL engine for quick reporting and *ad hoc* analysis. It follows parallel computing to overcome input and output latency. Presto provides an opportunity for users to create their own functions, thereby increasing user-friendliness.

6.4 INTRODUCTION TO MACHINE LEARNING

Over the past several years, a large amount of data has been created than in the millennia of history before. These data have a very high commercial value. Without proper tools and necessary processing, the tremendous amount of data becomes ineffectual to the users.

ML consists of various algorithms that allow application software to become extensively precise in predicting results without being distinctly programmed. The basic proposition of ML is to construct algorithms that can accept input data and use statistical interpretation to forecast a result while generating results as new details become accessible. The vital role of ML comprises self-learning algorithms that develop continuously by enhancing their designated task. When put together correctly and given proper data, these ML algorithms finally produce results in the format of pattern recognition and predictive modeling. In the case of ML algorithms, data are like practice; therefore, more practice makes it better to understand the data. Algorithms optimize themselves with the available data they train with similar to the method by which Olympic athletes sharpen their bodies and skills by preparing daily (Huljanah et al. 2019). Numerous programming languages work along with ML involving Python, R, Java, and JavaScript. Python is the popular choice for many developers because of its Tensor Flow library, which helps to give a comprehensive ecosystem of ML tools. ML can be used all over from industrializing mundane tasks to contributing intelligent apprehensions. Commercial enterprises from various fields can also try to benefit from it—for example, a home assistant like Google Home or a fitness tracker like Fitbit. Facial recognition software enables multimedia to help users share photos of their friends. Optical character recognition machinery transforms images of text into machine-encoded text. Most of these streaming platforms use recommendation engines that use collaborative filtering techniques which are powered by ML.

6.5 TOOLS USED IN MACHINE LEARNING

ML has become renowned and highly effective mainly because of a variety of tools available which include both open-source and paid tools (Hasan et al. 2016). The unique specialty of ML tools is a high degree of user-friendliness and effective execution of intended prediction at a swift rate. Following are the most popularly used ML tools which help us in performing various ML-related task with ease.

6.5.1 Tensor Flow

It provides an open-source JavaScript library that assists in ML developments. The application programming interface will enable us to create and instruct the models. In addition, it has a unique capability to work with multidimensional arrays efficiently. Similarly, its scalability is quite high in huge dataset computations.

6.5.2 Google Cloud

Google Cloud (GC) is an AI platform which is used for tension-free development. Moreover, GC is an organized program that allows ML developers and data scientists to build and process highly effective and accurate ML models. It is suitable for all who want faster results with automatic code generated. GC is optimal for data scientists of any skill level.

6.5.3 AWS ML

AWS is a cloud-based ML model that mainly focuses on learning software application suitable for software developers. AWS service is created also for multiple input predictions, and these predictions happen in an asynchronous manner. It also acts as an evaluation measure to test the quality of various ML techniques.

6.5.4 Accord and Apache Mahout

Accord is .net framework that is developed in C#. It is a repository for image and audio libraries. The accord's main function is to provide libraries for several real-time applications. Apache Mahout is an open-source tool which is highly suitable for mathematical operations and visualizations of different charts. Hence, it is a widely opted tool by various mathematicians. Some of its features include acceleration of GPUs followed by database diagnostics.

6.5.5 Shogun

It is an open-source module which is used in ML. It is programmed in C++. It provides algorithms for ML problems. It is mainly used by practitioners, hackers, scientists, and idealists. It has the capability to process up to ten million samples (Singh et al. 2016).

6.5.6 Oryx 2

It is built by a combination of Apache Kafka and Spark. It is mainly used for prediction using various ML algorithms on a real-time basis. It is mainly used for app development and creating several real-time applications for data filtering.

6.5.7 Apache Singa

It is broadly used in image identification and NLP. It provides a broad spectrum of various neural networks model. It consists of three main features which include input and output followed by CPU core and lastly the ML model.

6.5.8 Google ML Kit for Mobile

It can be used for face identification, text interpretation, and scanning bar code applications. Some of its unique features include segmentation of selfies, extraction of entities, and various poses detection.

6.5.9 Apple's Core ML

This model is highly streamlined for mobile applications. Core ML combines Apple hardware and various ML models increasing the device performance and maintaining battery life simultaneously.

6.6 BIG DATA TYPES AND ITS CLASSIFICATIONS

Big data can be broadly classified into three types, namely, structured, unstructured, and semi-structured. Each and every data format which is generated will fall into either of these categories. The detailed explanation of each type is given in subsequent sections.

6.6.1 Structured

Structured data is a type of big data, and this organized data can be refined, saved, and captured in a definite format. It refers to highly structured data that can be readily and seamlessly managed and can be retrieved from a database by simple search engine algorithms. For example, the employee table in an organized database will be organized as employee details that include employee name, employee ID, employee age, and so on followed by their job position and salaries which can be represented in a structured format (Ma et al. 2014).

6.6.2 Unstructured

Unstructured data implies that the data lacks any particular form or structure whatsoever. This, in turn, makes it a difficult and time-consuming procedure to evaluate the unstructured data. For instance, email is an example of unstructured data.

6.6.3 Semistructured

Semi-structured is the third type of big data. This type of data contains information organized in both structured and unstructured formats. This model refers to the data that has not yet been organized still comprises vital information. Big data can be classified into five different subcategories, namely, volume, variety, veracity, value, and velocity. These five Vs cover all the possible data.

6.6.4 Volume

This indicates the unimaginable amounts of data generated from various social media platforms, mobile phones, vehicles, credit cards, and images. At present, we are using a distributed system to contain the data in several locations, which is then combined together by a software framework called Hadoop.

6.6.5 Variety

Big data includes multiple varieties of data. Traditional data includes phone numbers and home addresses, whereas nowadays, a large amount of data is in the form of photos, videos, and much more, and all these are completely unstructured.

6.6.6 Veracity

It basically refers to the degree of accuracy the data has to offer. A huge portion of the data is unstructured; it needs to find a way to filter them or convert them into useful or meaningful data for business developments.

6.6.7 Value

Value is one of the key factors to concentrate on in terms of big data. Even if we have a lot of data, it should be valuable, reliable, and trustworthy which needs to be processed and analyzed to find insights.

6.6.8 Velocity

This is another key component for big data with a condition that unless the data is processed and analyzed and delivered on time, it is pointless to hold on to that data as it can get outdated (Passos et al. 2019).

6.7 LATEST TREND IN BIG DATA

It is a surprising fact that the data that we produce in two days is much higher than decades of history. Over the years, there has been a lot of transformations which have occurred, thereby making data science a prominent field. Further technological advancements will result in mass production of data. Following are a few trends observed in big data analytics (Karuppiah et al. 2014).

6.7.1 Information from Data

A few years back, data stores were repository for data. While SaaS was widely used, Daas had just started. Moreover, SaaS-driven applications focused upon utilizing cloud technology which aids clients using the application with on-demand access privilege irrespective of the location. The greatest feature that big data analytics provides is transforming raw data into information which can be interpreted. This helps develop businesses and share informative data among the industry (Tolan et al. 2015).

6.7.2 Predictive Analysis

Analysis of big data has been a pivotal factor which has helped companies to accomplish their agendas, thereby having greater success. They employ analytics tools to train big data and find out why issues arise. Predictive analytics plays a conducive role in analyzing data from various events which help in enhanced knowledge of customers leading to an accurate prediction of hazards the corporation is vulnerable to. Furthermore, analysis of big data helps in forecasting future happenings. This approach is extensively useful in analyzing customer's responses efficiently (Karuppiah et al. 2015).

6.7.3 Edge Computing

The edge computing works on the principle of transferring multiple processes to a local system or server and running it on them. This helps in reducing distant connections which interlink servers and customers. Furthermore, edge computing involves real-time streaming of data and analyzing the data without being affected by latency difficulties. This, in turn, helps to respond much faster. It is a structured way to process a large amount of data by taking in less bandwidth.

6.7.4 Natural Language Processing

NLP is a subdivision of AI that helps humans to communicate with systems in real time with ease. Major objective of NLP is to decode human-based emotions and language and further analyze it. ML plays an important role in processing human-understandable language to natural machine language. NLP uses various kinds of algorithms to extract data in compliance with grammatical rules which are predefined and analyze it further. The semantic technique is one of the most used methods in natural language processing (Dreiseitl et al. 2002).

6.7.5 Hybrid Clouds

This type of system utilizes a combination of public and private clouds with synchronization between two clouds. Increasing the data deployment option increases the adaptability among the private and public clouds. To build a hybrid cloud, the company must opt for the usage of a virtualization layer which helps virtual machines in installing private cloud. Using this software data can be passed between public and private clouds (Karuppiah et al. 2019).

6.7.6 Dark Data

Dark data are those data which are not suitable for usage in any systematic approach or system. Hence, they are highly discouraged by an organization to be used for any official task. The data which is obtained is not utilized to find insights or predictions which make it unique as compared to other data. Even though dark data cannot be explored due to lack of technical capabilities, companies still consider the scope of using them in the future (Boulesteix et al. 2012).

6.8 TYPES OF MACHINE LEARNING ALGORITHMS

In ML, there are broadly three main classifications as mentioned earlier. Figure 6.1 shows the broad classification and is explained in detail in the following sections.

6.8.1 Supervised Learning

In supervised learning, training of "labeled" data takes place. It means that along with various attributes of data, the outcome is also given. The main motive of supervised learning is to predict outcomes for unforeseen data. Supervised learning helps

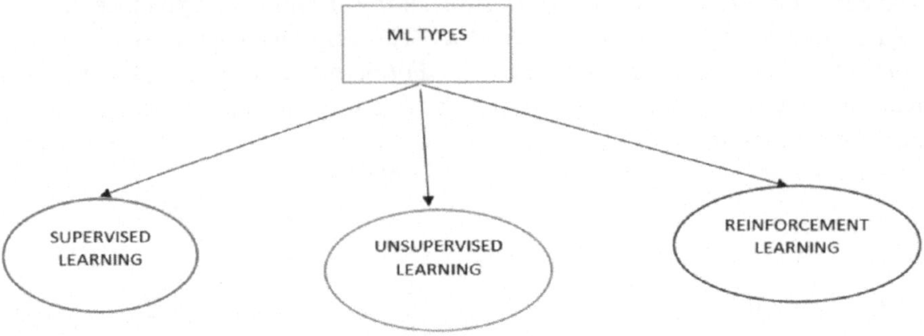

FIGURE 6.1 Machine-learning types.

to produce data from existing data. It aids in optimizing the performance criteria using training data. Supervised learning helps in solving various real-world issues (Chen et al. 2014).

6.8.1.1 Categories of Supervised Learning

6.8.1.1.1 Classification

Classification refers to distributing data from data points into various classes based on several mathematical functions. In classification, there exists a target or label that acts as an output variable. The main function of classification predictive modeling is to approximately map input variable to output variable (Wu et al. 2018). Let's take an example: spam detection is one of the common problems faced by everyone. The spam detection can be broadly classified into two different classes which include "spam" and "not spam." Hence, this falls under the classification category, and the output variable will contain two classes. Using appropriate classification algorithms, machines can be well trained and can predict accurately on testing, whether a given email is spam or not (Barbella et al. 2009). Table 6.1 shows classification types, and Table 6.2 shows classification algorithms.

6.8.1.1.2 Regression

Regression is mainly for predicting when output variables are continuous. It is a statistical process in which interrelationships between dependent and independent variables are focused. Regression covers domains—namely, finance and stocks (Punithavathi et al. 2019). Table 6.3 projects regression algorithms.

6.8.2 Unsupervised Learning

Here, the machine is trained with data which is unlabeled. There is no variable called "output variable or categories" in unsupervised learning. These algorithms are opted when the analyst doesn't know what the main outcome of the data is. Figure 6.2 shows the types of clustering algorithms. Unsupervised learning streamlines its

TABLE 6.1
Classification Types

S.No.	Classifier Type	Properties
1	Binary classification	In this type, only two possible outcomes are present. Example: Gender Classification (Male/Female)
2	Multiclass classification	This type of classification contains multiple categories such that a sample assigned to this classification will only belong to either of these categories and not both. Example: Animal can be a cat or a dog but not both at the same time.
3	Multilabel classification	In this type, each sample can be linked to more than one category. Example: A newspaper sample will contain columns of sports, politics, along with science and technologies.

TABLE 6.2
Classification Algorithms

S.No.	Algorithm Name	When It Can Be Used	Advantages	Disadvantages
1	Logistic regression	i. It is useful to predict the outcome of variable which is categorical type data from predictor variables, which may be continuous or categorical. ii. It is mainly used because categorical outputs violate the linearity assumptions in normal regression. iii. Logistic regression does the logarithmic transformation of output variable which allows to model nonlinear type in a linear manner.	i. Very easy to understand and implement. Furthermore, very much efficient to train. ii. There are no assumptions made about the distribution of categories among the features. iii. High-accuracy output will be obtained for simple and linearly separable datasets.	i. Linear boundary construction. ii. Possibility of overfitting if the number of observations is less than the number of features. iii. Linearity assumption between independent and dependent variables.
2	Naïve Bayes	i. Works well when multiple classes are present. ii. Suitable for classification of text. iii. Optimal only for dataset with smaller dimensions (rows and columns).	i. Faster processing and lesser time complexity. ii. Highly suggested for prediction of multiclass problems. iii. If there is independence of data, then it is the best classification model.	i. If features are not independent, then not an efficient algorithm. ii. If there is a category which is present in test set and not in train set, then Naïve Bayes will fail to make predictions (zero frequency). iii. Often a lousy estimator.
3	Stochastic gradient descent	i. Efficient model which is used to fit linear models in an easier manner. ii. Highly recommended when the sample size is very large. iii. Includes various loss functions and penalties for classification.	i. Less memory consumption. ii. Very fast in computation as processing of samples takes place one by one. iii. Suitable for large datasets since parameter updating occurs frequently	i. May lead to change the gradient descent alignment due to frequent updates being very noisy. ii. Deals only with a single example at a time. iii. Computationally expensive

4	K-nearest neighbors (KNN)	i. Less dimensions of dataset. ii. When noise-free data is available. iii. If data is properly labeled data.	i. Implementation is very easy. ii. New data can be added at ease. iii. It learns during the time of real predictions.	i. Won't get optimal results for a large dataset. ii. Not suitable with higher dimensions. iii. Accuracy will be greatly affected if there are noisy data, missing values, and outliers.
5	Decision tree	i. It can be used for cases where a sequence of rules can be used to classify the data in a tree structure.	i. Normalization of data isn't needed. ii. No scaling of data is required. iii. Easy mode of explanation to stakeholders.	i. Time consumption for training is high. ii. Complex calculation as compared to the majority of algorithms. iii. Not suitable for regression and continuous values predictions.
6	Random forest	i. Less sensitive to overfitting. ii. When highly accurate results are targeted, random forest is a must. iii. Random forest is the enhancement of decision tree, and hence it is easily interpretable.	i. Reduced overfitting, hence highly accurate. ii. No normalization is required. iii. Missing values will be filled automatically.	i. Requires high computational power. ii. Not suitable for regression and continuous values predictions. iii. Time consumption for training is high.
7	Support vector machine (SVM)	i. SVM can be used if memory efficiency is a prime objective. ii. When there is a large dataset, SVM can be relied upon.	i. Provides high accuracy when there is a clear distinctive margin. ii. Larger dimension dataset can be easily handled by SVM. iii. Highly memory efficient.	i. Training time might be high, depending on dataset. ii. Noise-sensitive.

TABLE 6.3
Regression Algorithms

S.No.	Algorithm Name	When It Can Be Used	Advantages	Disadvantages
1	Linear regression	i. When the degree of outliers is less. ii. When you want to easily model the data and interpret the data. iii. Useful when their size of the dataset is small.	i. Easy to model and interpret. ii. Is susceptible to overfitting. iii. Less complex as compared to other algorithms.	i. Accuracy is greatly affected by outliers. ii. Linear regression does not offer the complete relation among variables. iii. It works on the assumption of independence of attributes.
2	Polynomial regression	i. Useful to model data which is nonlinearly separable. ii. When complex relationships are present, then polynomial regression can be used. iii. Suitable when exponents are selected carefully.	i. Provides highly accurate predictions irrespective of the size of data. ii. Highly suitable even for nonlinear problems.	i. Poor selection of exponents can lead to overfitting. ii. Careful choosing of the right polynomial degree is needed for good bias.
3	Ridge regression	i. Suitable when normality is not assumed.	i. Overfitting is prevented. ii. Compromises variances for bias.	i. Bias is increased. ii. Hyperparameter selection is difficult.
4	Lasso regression	i. When built-in feature selection is needed. ii. When computational efficiency is required.	i. Overfitting is prevented. ii. Has built-in feature selection ability.	i. Bias is increased. ii. Accuracy is less as compared to ridge regression.
5	Elastic net regression	i. When there is no limit to a number of selected variables. ii. On highly correlated variables, it focuses on group effect.	i. No limit to a number of selected variables.	i. More time consumption than lasso and ridge regression.
6	Logistic regression	i. It is useful to predict the outcome of variable which is categorical type data from predictor variables, which may be continuous or categorical. ii. It is mainly used because categorical outputs violate the linearity assumptions in normal regression. iii. Logistic regression does logarithmic transformation of output variable which allows to model nonlinear type in a linear manner.	i. Very easy to understand and implement. Furthermore, very much efficient to train. ii. There are no assumptions made about the distribution of categories among the features. iii. High-accuracy output will be obtained for simple and linearly separable datasets.	i. Linear boundary construction. ii. Possibility of overfitting if the number of observations is less than the number of features. iii. Linearity assumption between independent and dependent variables.

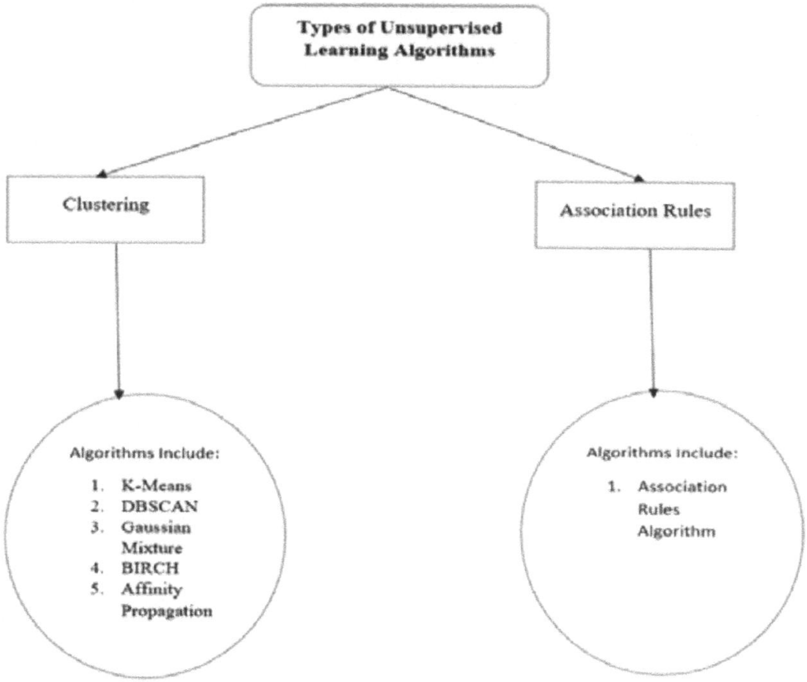

FIGURE 6.2 Types of unsupervised learning algorithms.

functionalities to pattern detection and descriptive modeling. Furthermore, the data is summarized for providing meaningful insights into the data.

6.8.3 Reinforcement Learning

Reinforcement learning algorithm is often referred as an agent. An agent follows an iterative approach in learning from the environment. This algorithm is designed in a manner where the agent receives both reward and penalty for the action it takes, thereby enabling the agent to explore several states. The greatest advantage of using reinforcement learning is the ability of software agents in identifying the best possible action for a given task, thereby maximizing the performance in the shortest time span possible. One of the star features of reinforcement learning is the reward feedback obtained which helps theagent learn about its actions (Xhemali et al. 2009).

6.8.3.1 Reinforcement Learning Steps

1. Agent observes the initial state.
2. Decision-making function is defined based on which agent performs an action.
3. Based on the action performed, the agent receives reward or reinforcement.

4. Each action and its state followed by reward status are stored.

Common Reinforcement-Learning Algorithms: DAN, TD, and Q-learning
Reinforcement-Learning Applications: Self-driving cars, computer chess, robotic hands, and so on.

6.8.4 Types of Reinforcement Learning

1. *Positive*
   Positive reinforcement occurs when a particular action taken by the agent creates a positive impact, thereby maximizing performance capability.

2. *Negative*
   Negative reinforcement occurs when a particular action taken by the agent creates a negative impact, thereby decreasing performance capability. It is known for strengthening behavior as the agent learns to avoid it when the condition repeats again.

6.9 GUIDELINES ON OPTIMAL STEPS IN MAKING MACHINE LEARNING PREDICTIONS

There is a series of steps which need to be executed, which affect the accuracy of prediction using various ML algorithms. Each of these individual methods has a significance of its own. Furthermore, it helps in understanding the dataset better, which gives insights on which algorithms to use and what factors contribute most to the prediction (Tian et al. 2015). At the end, visualization will be easier since sound knowledge of dataset is obtained; optimal graphical representation can be chosen as well. Figure 6.3 shows the suitable method for accurate prediction using ML. Given later is a detailed explanation of the most important factors that need to be considered while doing a prediction (Bertsimas et al. 2020).

6.9.1 Data Selection

Choosing the appropriate factors for prediction along with sufficient data collected from relevant sources is the first step to be done. The accuracy of predictions hugely depends on the type of dataset and the content of dataset. Hence, data selection is an important factor.

6.9.2 Data Exploration

Exploring the data from dataset is the second step that needs to be executed. It includes summarizing the data with various different mathematical measures such as mean, median, and standard deviation followed by graphical visualization for gaining insights on how the data is distributed. Similarly, correlation between different factors in dataset is analyzed in this step as well (Zhu et al. 2021).

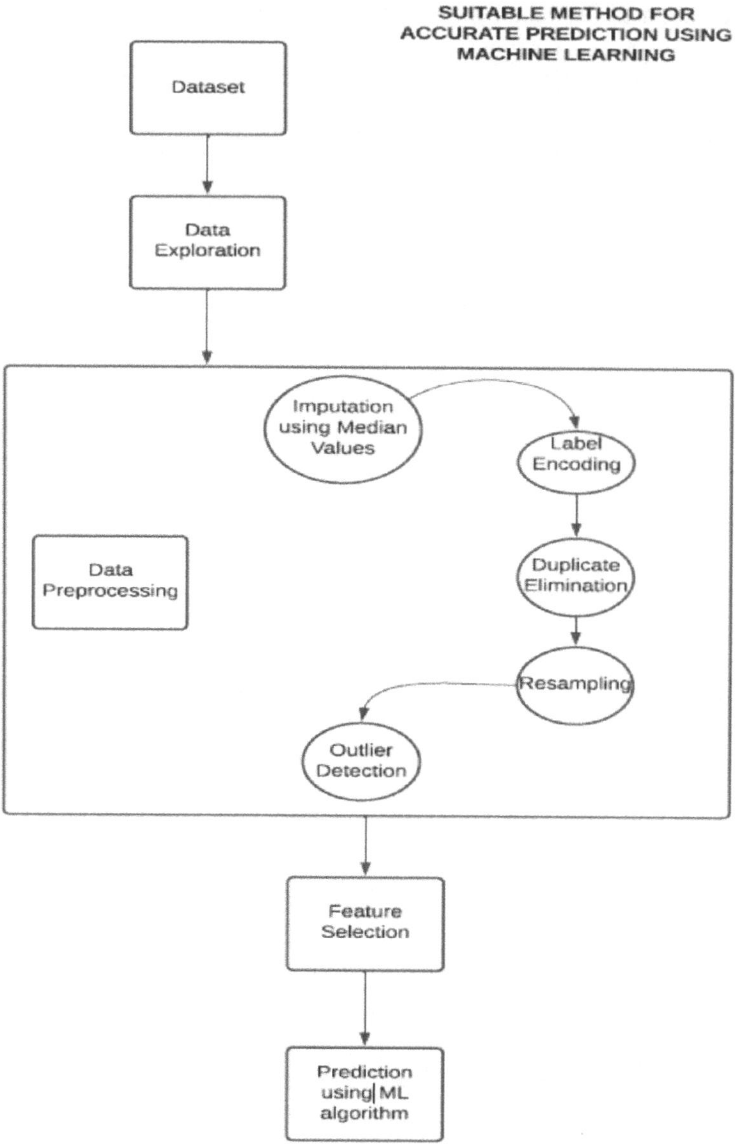

FIGURE 6.3 Suitable method for accurate prediction using machine learning.

6.9.3 Data Preprocessing

Refining of data is often referred as data preprocessing. It includes various steps. First is handling missing values. A few common techniques used are list-wise deletion, pairwise deletion, or imputation using median value with the latter being the most optimal method. In imputation using median values, wherever null values are obtained, the particular column's mean value is calculated and entered in the missing place. The next important step is label encoding. Label encoding is mostly needed when dataset has an output variable, and it is a string value (e.g., yes or no). The most optimal method is to convert each of the words to a numeric value, thereby making data suitable for training. There are possibilities of having duplicate entries in dataset. This might cause incorrect predictions. Hence, it is a must to remove duplicate entries. Furthermore, resampling of data might come in handy when there is not sufficient data present for making accurate predictions. Resampling techniques are used for increasing or decreasing the amount of data. To maximize data, several algorithms such as SMOTE and boruta learn from data present and then maximize the dataset. In case of minimizing data, data which is largely deviated from central tendency will be removed (Jan et al. 2019). The last step that needs to be checked is the presence of outliers. Outliers can result in incorrect result. An outlier can be predicted graphically using a box-plot. There are several algorithms to identify the outliers—for example, isolation forest. All these steps help in getting refined data, which helps in making accurate predictions.

6.9.4 Feature Selection

Selecting appropriate features from dataset is very important for predictions. Removing those attributes doesn't contribute to the model which may negatively impact the performance. Various algorithms are used for selecting appropriate features. One of the most prominent algorithms is genetic algorithm. Genetic algorithm is inspired by Darwin's method of natural selection. Genetic algorithm follows initialization, fitness assignment, selection, cross-over, and mutation. Figure 6.3 projects suitable method for accurate prediction using ML.

6.9.5 Training and Testing

The first task is to divide your entire dataset into training and testing dataset. Conventionally, it is divided as 80% training set and 20% testing set or 70% training and 30% testing. The major reason for doing so is that the machine will be learning about the data and the correlation among the data based on this training set. Machines' capability to predict outcomes is tested via the testing set. This division helps in predicting information for unforeseen data (Chen et al. 2014).

6.9.6 Prediction Using Various ML Algorithms

Once the training and testing data frames are created, selecting the right algorithms aids in getting accurate results. If there is an output variable available, then supervised ML algorithms such as logistic regression and KNN can be used. Also, bagging algorithms such as decision tree, random forest, light GBM, and stack estimator can also be used for predictions (Bos et al. 2014). In case there is no target variable available, then unsupervised learning is the best option. Algorithms such as polynomial regression, lasso regression, and logistic regression can be used. Furthermore, multiple algorithms of the type chosen can be used for understanding how each algorithm predicts and which algorithms provide accurate predictions (Landset et al. 2015).

6.10 BIG DATA ANALYTICS AND MACHINE LEARNING FUSION

To predict results accurately, several varieties of data are important factors. The biggest obstacle which arises is regarding the management of these wide varieties of data. Moreover, the information gathered without defined analysis and interpretation is useless. This problem is resolved in fusion with ML. ML algorithms enable providing meaningful information from big data (L'heureux et al. 2017). Hence, efficient business operations and improved service quality can be achieved. Big data provides enormous varieties of data. ML acts as a back end, thereby receiving the data and performing training and testing using suitable algorithms. At the end, the machine can accurately predict the results with lesser time complexity. Fusion of big data and ML has opened doors to the faster computation of different data (Qiu et al. 2016).

6.11 ADVANTAGE OF BIG DATA AND MACHINE LEARNING

6.11.1 Analyzing Data in a Limited Time Frame

ML techniques can be used to process an unlimited amount of data and provide appropriate predictions for the raw data collected. Hence, ML has become the x-factor which drives businesses to huge profits and helps them in reaching greater heights. ML also possesses the capability of forging data from multiple sources, thereby providing accurate results efficiently. All this processing of ML only depend on the dataset. Still, it's a highly optimal method as compared to its conventional counterparts (Pääkkönen et al. 2015).

6.11.2 Prediction of Real-Time Data

Data scientist considers ML as the most efficient technique for executing meaningful predictions. Businesses from various domains depend on ML techniques for faster analysis of the company's performance and forecast the future of the company. Few examples include stock market prediction, sentiment analysis, and liver disease prediction. The combination of big data and ML helps data analyst to deepen their research and get meaningful analysis out of the big data collected (Najafabadi et al. 2015).

The merging of big data and ML helps identify loopholes that could have driven potential prospects away. Analysis of various patterns, history of revenue, and so on help data scientists plan out their objectives and the type of customers to target. This will further guide to use various marketing tools to gain the customers (Lawless et al. 2010). McDonalds is a suitable example for demonstrating the success of ML and big data. Based on the feedback received from various customers, they analyze which age groups prefer eating their product and what is it that customers are expecting from them. Hence, it helps Mc Donalds to determine what new strategies must be implemented to reach out to more customers (Hadioui et al. 2017).

6.12 TRADE-OFF OF BID DATA AND ML COMBINATION

6.12.1 Data Acquisition

Big data which is collected from various sources may be unfiltered, and, hence, there are high possibilities for high bias. Hence, several preprocessing needs to be done before using ML algorithms. This task can be tedious depending on the size of data (Qiu et al. 2016).

6.12.2 Time and Resources

Depending upon the machine algorithms, there might be a drastic difference in the learning rate of data. Further on increasing data size, learning time can be longer as well.

6.12.3 High Error Susceptibility

There is a high probability of a chain of errors to go undetected when prediction occurs. Even if it gets detected, it is extremely difficult to find the root cause.

6.13 APPLICATIONS OF BIG DATA AND MACHINE LEARNING

On a daily basis, tons of data are being produced and are stored at various data stores. We also know that companies use these data in order to get insights into more valuable information. But if we are not able to train data in such a way that it can give a meaningful value to companies, then these stored data is of no value to them. Hence, both big data and ML can be combined, and the results can be used in various fields as shown in the following sections.Following are a few real-life examples that show the application of ML and big data (Istepanian et al. 2018).

6.13.1 Cloud Networks

For a firm to train an enormous load of data on-site can be a challenge as it requires a lot of servers, storage space security frameworks, and much more, which all lead to a huge amount of money. Amazon EMR is a cloud platform that gives us various data analysis models, all of them confined in a single framework. Such ML models possess features such as text classifications and image recognition. The major drawback of these is the lack of deployment and limited support (Khanzode et al. 2020).

6.13.2 Web Scraping

A manufacturer needs to understand the market trends and customer satisfaction to make the products more suitable to the customers. So, he decides to scrape all the responses given for his product from the website. After segregation of the obtained data, it is then put under various ML models in order to train it and understand how to modify the products which can help in increased sales. As we get a lot of data after processing us, filtering those data as aggregation is one of the most important processes involved in this.

6.13.3 Mixed-Initiative Systems

Netflix recommendation system works on the principle of collaborative filtering. Here, the history of shows of each user can be considered as big data, and using this history ML models can predict which all types of shows the user is interested in. This sample shows how ML and big data are closely interlinked. A similar method is used in making smart cars as those decisions are being generated by the user after all the previous data that has been procured in the past (Dina et al. 2013).

6.14 GUIDELINES ON HOW MACHINE LEARNING CAN BE EFFECTIVELY APPLIED TO BIG DATA

The working of combination of big data and ML is contrasting to regular ML. While there is a wide range of ML algorithms for predictive analysis, the performance of the prediction is not solely dependent on ML algorithms; the ability to transform data suitable for ML task is a must for effective analysis. Following are the best possible guidelines in effectively applying ML to big data (Saravanan et al. 2018). Figure 6.4 shows guidelines on how ML can be effectively applied to big data.

6.14.1 Data Warehousing

Data warehousing is the preliminary step to be conducted. It is a process of collecting data from several sources and managing data to provide business insights. It is a fusion of components and the latest technology which helps in an effective use of data. There are three major classifications of data warehouses, namely, enterprise data warehouse, operational data store, and data mart. This is the right option for a strategist who depends on a lot of data. Offline operation database, offline data warehouse, real-time data warehouse, and integrated data warehouse are the general stages of data warehousing. It is the starting step in discovering patterns of groupings and data flows (Zhou et al. 2017).

6.14.2 Data Segmentation

Segmentation of data is a mechanism of splitting the data obtained followed by amassing alike data together. The grouping of data is based on certain predefined parameters. Data segmentation is an important aspect that needs to be focused because it reduces the computation cost and the analysis complexity of big data (L'heureux et al. 2017). Moreover, domain-specific messages can be made, ones which attract the customers better. To effectively apply data segmentation, highly accurate data and effective data quality tools are inevitable. Few types of data that provide great insights

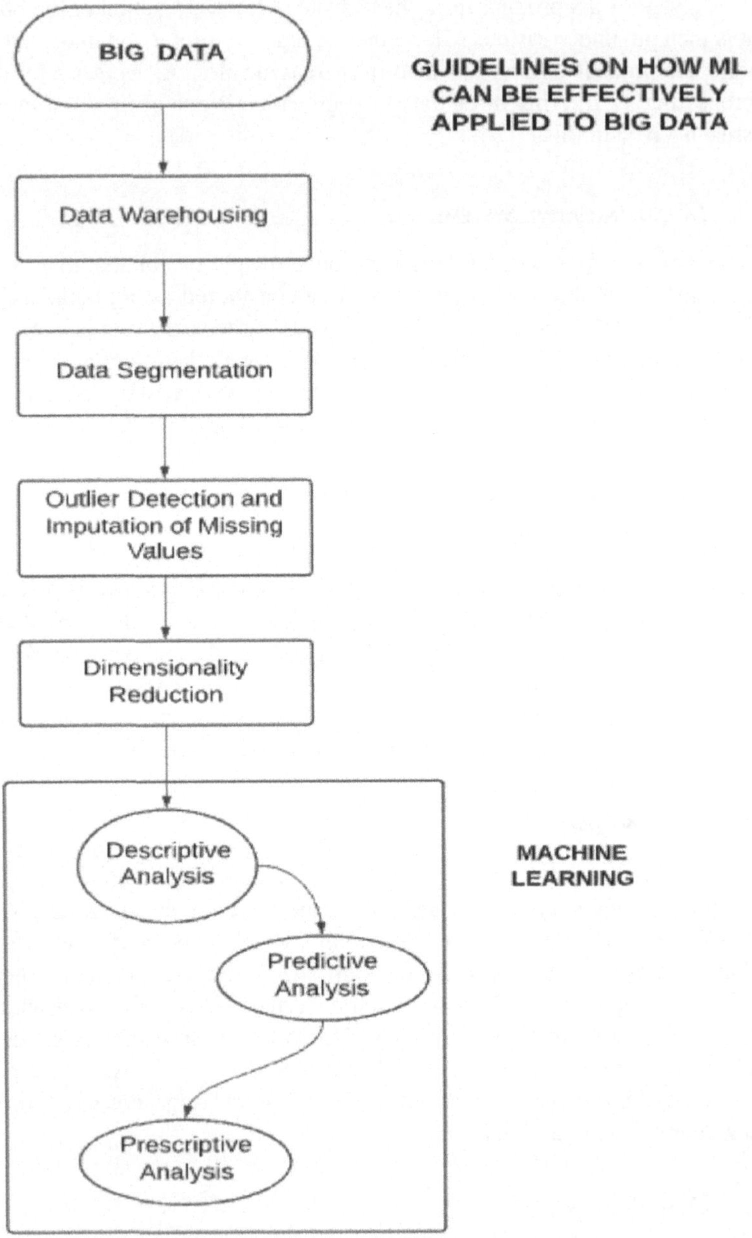

FIGURE 6.4 Guidelines on how ML can be effectively applied to big data.

are namely transactional data and behavioral data. Transactional data includes various information such as orders count, date of order, products, and categories of products. Behavioral data includes sessions count, session timings, and so on (Ngiam et al. 2019).

6.14.3 Outlier Detection and Imputation of Missing Values

The presence of outliers and missing values disrupt the extraction of useful information from the data. Furthermore, the lack of data quality is being indicated by the presence of outliers and missing values. Outliers denote extreme data points from the normally distributed data. Detection and removal of outliers followed by an elimination or imputation of missing values is a pivotal task. The most prominent outlier detection techniques are DBSCAN, isolation forest, Z-score, local outlier factor, and numeric outlier detection. Whereas for imputation of missing values, the following techniques are useful, namely, mean imputation, hot-deck imputation, cold-deck imputation, regression imputation, stochastic regression imputation, interpolation, and extrapolation. By doing so, cleansing of data is achieved (Chong et al. 2015).

6.14.4 Dimensionality Reduction

Visualization, computation time, and space complexity are undermined by the number of features present in the data. Most of the features present are highly correlated with each other due to which using all these features for prediction using ML is unessential. Dimensionality reduction helps in identifying the principal features (highly important features) and eliminating other unwanted features. Dimensionality reduction includes feature selection and feature extraction as its components. Feature selection includes filtering, wrapping, and then, at the end, embedding the useful features (Priyanka et al. 2014). Whereas, feature extraction is converting data from higher to lower dimensions. A few prominent methods of dimensionality reduction are PCA, LDA, and generalized discriminant analysis. Furthermore, dimensionality reduction can switch between both linear and nonlinear methods. Linear methods are, namely, PCA, factor analysis, and LDA. Nonlinear methods include Kernel PCA, Isomap, t-distributed stochastic neighbor embedding, and multidimensional scaling. PCA is one of the most famous linear methods. Following steps are involved in PCA; at first, the covariance matrix needs to be constructed (Van Der Maaten et al. 2009). Next eigenvectors are computed followed by largest eigenvalue, and its corresponding eigenvector is used to formulate a fraction of the original data. Performing dimensionality reduction helps in saving a lot of storage space and computation time (Singh et al. 2019). Data loss and a lack of efficiency in the reduction of variables when variables are linearly correlated are few hazards that need to be taken care of while performing PCA. To keep the most important features, backward elimination, forward selection, and random forest technique are suitable (Manogaran et al. 2018).

6.14.5 Descriptive Analysis

Descriptive analysis is one of the techniques used to transform raw data into interpretable and meaningful data. This type of analysis streamlines in describing the data both statistically and graphically. Visualization of various plots based on the data helps to gain more insights about the data. Statistical analysis helps in understanding the data distribution along with the detecting anomalies in data. Descriptive analysis is broadly classified into two subcategories: data mining and data aggregation. Data aggregation involves collection and sorting of data making it more viable. Data mining focuses on manipulating and refining data, thereby making it suitable for further analysis. Several measures such as inequality, discrimination, and segregation can be discovered using descriptive analysis (Sughasiny et al. 2018). Different types of descriptive analysis are measure of frequency, measure of central tendency, measure of dispersion, measure of position, contingency tables, and scatter plots. Measure of frequency is important to find out count or percentage of the attribute. Measure of central tendency is an excellent metric used to identify mid values. It involves mean, median, and mode. Measure of dispersion is needed to figure out how the data is dispersed. It includes standard deviation, variance, and coefficient of variation and range. Scatter plots, bar charts, column charts, and bubble charts are other visualization tools used for describing the data (Rahul et al. 2021).

6.14.6 Predictive Analysis

The next step to be followed after descriptive analysis is predictive analysis. The combination of ML techniques, statistical algorithms, and data are used to forecast future outcomes. Analysis of big data has been a pivotal factor that has helped companies to accomplish their agendas, thereby having greater success (Xing et al. 2015). They employ analytics tools to train big data and find out why issues arise. Predictive analytics plays a conducive role in analyzing data from various events, which helps in enhancing the knowledge of customers leading to accurate prediction of hazards the corporation is vulnerable to. Furthermore, analysis of big data helps in forecasting future happenings. This approach is extensively useful in analyzing customer's response efficiently. Predictive analysis is utilized for various activities such as fraud detection, risk reduction, and contingency planning for possible upcoming risks in the future (Santos et al. 2016). Furthermore, it is used for upgrading the business operations and market campaign optimization. There are several industries that use predictive analysis, namely, retail industry, banking sector, government sector, health industry, and manufacturing industry. The major advantages of predictive analysis are identifying the customer satisfaction rate and planning to increase the rate. Moreover, it helps in finding the target customers who will help in gaining more profit for the company. Furthermore, predictive analysis helps in identifying the purchase probability of a product (He et al. 2016).

6.14.7 Prescriptive Analysis

Prescriptive analysis is the concluding method in business intelligence. It is the next step after predictive analysis. This type of analysis does not only restrict itself to forecasting about the future but also considers the possible consequences and impacts based on the future as well. Prescriptive analysis focuses on how certain strategies can be made to occur in the future. It is not restricted based on any kind of rules which are static. Prescriptive analysis is purely based on the data which is gathered and the insights gained from the previous steps (Xing et al. 2016). Since it's purely based on data, hence several biases based on personal opinions and intuitions are completely discarded. Moreover, this type of analysis provides tangible and realistic measures that provide business solutions based on the business domain. It is quite complex to do prescriptive analysis, and it requires a team with highly skilled data scientists. Major skills that are required are mainly a team that is good at configuration of the analytics model and, in addition, a team who can clearly state the business problem for which solution is searched. Furthermore, a team that is able to mix analysis with business preparation and planning. Few prescriptive analytics software vendors who are currently dominant in the market are SAS, IBM, FICO, and River Logic (Jesmeen et al. 2018).

6.15 CONCLUSION AND FUTURE WORK

In this big data-driven world, ML empowers businesses to unimaginable heights raising the standards of living. This chapter focuses on ML and big data concepts individually. Furthermore, fusion of ML and big data is explained. The chapter also discusses about advantages and trade-off of this fusion. Furthermore, guidelines for effective ML predictions and guidelines in effectively applying ML to big data are being clearly explained with the flowcharts. Big data and ML fusion is the best combination at present and will remain prominent in the future as well. Big data analytics and ML are key factors driving the global market. The world is currently big data driven, and big data will increase in importance at the future as well. A few significant changes that will probably occur are the following. Volume of data will increase exponentially from yottabyte to zettabytes within 5 years. Sophistication of ML will take place drastically which affects the information gathering from big data. Moreover, data science field will have more scope and will be in dire need of employees. Furthermore, privacy might be affected by the increasing data. Cyberattacks and cybercrimes will be unimaginable. At the end, each and every human need will happen at the forefront with small touch or thought.

REFERENCES

Barbella, David, Sami Benzaid, Janara M. Christensen, Bret Jackson, X. Victor Qin, and David R. Musicant. "Understanding support vector machine classifications via a recommender system-like approach." *DMIN*, vol. 1, pg. 305–311, 2009.

Bertsimas, Dimitris, and Nathan Kallus. "From predictive to prescriptive analytics." *Management Science*, vol. 66, no. 3, pg. 1025–1044, 2020.

Bos, Joppe W., Kristin Lauter, and Michael Naehrig. "Private predictive analysis on encrypted medical data." *Journal of Biomedical Informatics*, vol. 50, pg. 234–243, 2014.

Boulesteix, Anne-Laure, Silke Janitza, Jochen Kruppa, and Inke R. König. "Overview of random forest methodology and practical guidance with emphasis on computational biology and bioinformatics." *Wiley Interdisciplinary Reviews: Data Mining and Knowledge Discovery*, vol. 2, no. 6, pg. 493–507, 2012.

Chen, Weitao, Xianju Li, Yanxin Wang, Gang Chen, and Shengwei Liu. "Forested landslide detection using LiDAR data and the random forest algorithm: a case study of the Three Gorges, China." *Remote Sensing of Environment*, vol. 152, pg. 291–301, 2014.

Chen, Xue-Wen, and Xiaotong Lin. "Big data deep learning: challenges and perspectives." *IEEE*, vol. 2, pg. 514–525, 2014.

Chong, Dazhi, and Hui Shi. "Big data analytics: a literature review." *Journal of Management Analytics*, vol. 2, no. 3, pg. 175–201, 2015.

Dina, Aurora-Tatiana, and Silvia-Ileana Ciornei. "The advantages and disadvantages of computer assisted language learning and teaching for foreign languages." *Procedia-Social and Behavioral Sciences*, vol. 76, pg. 248–252, 2013.

Dreiseitl, Stephan, and Lucila Ohno-Machado. "Logistic regression and artificial neural network classification models: a methodology review." *Journal of Biomedical Informatics*, vol. 35, no. 5–6, pg. 352–359, 2002.

Fong, Simon, Suash Deb, and Xin-she Yang. "How meta-heuristic algorithms contribute to deep learning in the hype of big data analytics." In *Progress in Intelligent Computing Techniques: Theory, Practice, and Applications*, vol. 518, pg. 3–25. Springer, Singapore, 2018.

Hadioui, Abdelladim, Yassine Benjelloun Touimi, and Samir Bennani. "Machine learning based on big data extraction of massive educational knowledge." *International Journal of Emerging Technologies in Learning (iJET)*, vol. 12, no. 11, pg. 151–167, 2017.

Hasan, Md Al Mehedi, Mohammed Nasser, Shamim Ahmad, and Khademul Islam Molla. "Feature selection for intrusion detection using random forest." *Journal of Information Security*, vol. 7, no. 3, pg. 129–140, 2016.

He, Ying, Fei Richard Yu, Nan Zhao, Hongxi Yin, Haipeng Yao, and Robert C. Qiu. "Big data analytics in mobile cellular networks." *IEEE*, vol. 4 pg. 1985–1996, 2016.

Huljanah, Mia, Zuherman Rustam, Suarsih Utama, and Titin Siswantining. "Feature selection using random forest classifier for predicting prostate cancer." In *IOP Conference Series: Materials Science and Engineering*, vol. 546, no. 5, pg. 052031. IOP Publishing, 2019.

Istepanian, Robert S.H., and Turki Al-Anzi. "M-Health 2.0: new perspectives on mobile health, machine learning and big data analytics." *Methods*, vol. 151, pg. 34–40, 2018.

Jan, Bilal, Haleem Farman, Murad Khan, Muhammad Imran, Ihtesham Ul Islam, Awais Ahmad, Shaukat Ali, and Gwanggil Jeon. "Deep learning in big data analytics: a comparative study." *Computers & Electrical Engineering*, vol. 75, pg. 275–287, 2019.

Jesmeen, M.Z.H., J. Hossen, S. Sayeed, C.K. Ho, K. Tawsif, Armanur Rahman, and E.M.H. Arif. "A survey on cleaning dirty data using machine learning paradigm for big data analytics." *Indonesian Journal of Electrical Engineering and Computer Science*, vol. 3, pg. 1234–1243, 2018.

Kambatla, Karthik, Giorgos Kollias, Vipin Kumar, and Ananth Grama. "Trends in big data analytics." *Journal of Parallel and Distributed Computing*, vol. 74, no. 7, pg. 2561–2573, 2014.

Karuppiah, M., A.K. Das, X. Li, S. Kumari, F. Wu, S.A. Chaudhry, and R. Niranchana. "Secure remote user mutual authentication scheme with key agreement for cloud environment." *Mobile Networks and Applications*, vol. 24, no. 3, pg. 1046–1062, 2019.

Karuppiah, M., and R. Saravanan. "A secure remote user mutual authentication scheme using smart cards." *Journal of Information Security and Applications*, vol. 19, no. 4–5, pg. 282–294, 2014.

Karuppiah, M., and R. Saravanan. "A secure authentication scheme with user anonymity for roaming service in global mobility networks." *Wireless Personal Communications*, vol. 84, no. 3, pg. 2055–2078, 2015.

Khanzode, Ku Chhaya A., and Ravindra D. Sarode. "Advantages and disadvantages of artificial intelligence and machine learning: a literature review." *International Journal of Library & Information Science (IJLIS)*, vol. 9, no. 1, pg. 30–36, 2020.

Landset, Sara, Taghi M. Khoshgoftaar, Aaron N. Richter, and Tawfiq Hasanin. "A survey of open source tools for machine learning with big data in the Hadoop ecosystem." *Journal of Big Data*, vol. 2, no. 1, pg. 1–36, 2015.

Lawless, Harry T., and Hildegarde Heymann. "Descriptive analysis." In *Sensory Evaluation of Food*, pg. 227–257. Springer, New York, NY, 2010.

L'heureux, Alexandra, Katarina Grolinger, Hany F. Elyamany, and Miriam A.M. Capretz. "Machine learning with big data: challenges and approaches." *IEEE Access*, vol. 5, pg. 7776–7797, 2017.

Ma, Chuang, Hao Helen Zhang, and Xiangfeng Wang. "Machine learning for big data analytics in plants." *Trends in Plant Science*, vol. 19, no. 12, pg. 798–808, 2014.

Manogaran, Gunasekaran, and Daphne Lopez. "Health data analytics using scalable logistic regression with stochastic gradient descent." *International Journal of Advanced Intelligence Paradigms*, vol. 10, pg. 118–132, 1–2, 2018.

Manogaran, Gunasekaran, V. Vijayakumar, R. Varatharajan, Priyan Malarvizhi Kumar, Revathi Sundarasekar, and Ching-Hsien Hsu. "Machine learning based big data processing framework for cancer diagnosis using hidden Markov model and GM clustering." *Wireless Personal Communications*, vol. 102, pg. 2099–2116, 2018.

Najafabadi, Maryam M., Flavio Villanustre, Taghi M. Khoshgoftaar, Naeem Seliya, Randall Wald, and Edin Muharemagic. "Deep learning applications and challenges in big data analytics." *Journal of Big Data*, vol. 2, no. 1, pg. 1–21, 2015.

Ngiam, Kee Yuan, and Wei Khor. "Big data and machine learning algorithms for health-care delivery." *The Lancet Oncology*, vol. 20, no. 5, pg. e262–e273, 2019.

Pääkkönen, Pekka, and Daniel Pakkala. "Reference architecture and classification of technologies, products and services for big data systems." *Big Data Research*, vol. 2, no. 4, pg. 166–186, 2015.

Passos, Ives C., Pedro L. Ballester, Rodrigo C. Barros, Diego Librenza-Garcia, Benson Mwangi, Boris Birmaher, Elisa Brietzke et al. "Machine learning and big data analytics in bipolar disorder: a position paper from the International Society for Bipolar Disorders Big Data Task Force." *Bipolar Disorders*, vol. 21, no. 7, pg. 582–594, 2019.

Priyanka, K., and Nagarathna Kulennavar. "A survey on big data analytics in health care." *International Journal of Computer Science and Information Technologies*, vol. 5, no. 4, pg. 5865–5868, 2014.

Punithavathi, P., S. Geetha, M. Karuppiah, S.H. Islam, M.M. Hassan, and K.K.R. Choo. "A lightweight machine learning-based authentication framework for smart IoT devices." *Information Sciences*, vol. 484, pg. 255–268, 2019.

Qiu, Junfei, Qihui Wu, Guoru Ding, Yuhua Xu, and Shuo Feng. "A survey of machine learning for big data processing." *EURASIP Journal on Advances in Signal Processing*, vol. 1, pg. 1–16, 2016.

Rahul, Kumar, Rohitash Kumar Banyal, Puneet Goswami, and Vijay Kumar. "Machine learning algorithms for big data analytics." In *Computational Methods and Data Engineering*, pg. 359–367. Springer, Singapore, 2021.

Santos, Maribel Yasmina, and Carlos Costa. "Data warehousing in big data: from multidimensional to tabular data models." In *Proceedings of the Ninth International C* Conference on Computer Science & Software Engineering*, pg. 51–60, 2016, Association for Computing Machinery, New York, NY and Porto Portugal.

Saravanan, R., and Pothula Sujatha. "A state of art techniques on machine learning algorithms: a perspective of supervised learning approaches in data classification." In *2018 Second International Conference on Intelligent Computing and Control Systems (ICICCS)*, pg. 945–949. IEEE, 2018.

Singh, Amanpreet, Narina Thakur, and Aakanksha Sharma. "A review of supervised machine learning algorithms." In *2016 3rd International Conference on Computing for Sustainable Global Development (INDIACom)*, vol. 1, pg. 1310–1315. Ieee, 2016.

Singh, Sanjay Kumar, and Abdul-Nasser El-Kassar. "Role of big data analytics in developing sustainable capabilities." *Journal of Cleaner Production*, vol. 213, pg. 1264–1273, 2019.

Sughasiny, M., and J. Rajeshwari. "Application of machine learning techniques, big data analytics in health care sector—a literature survey." In *2018 2nd International Conference on I-SMAC (IoT in Social, Mobile, Analytics and Cloud)(I-SMAC) I-SMAC (IoT in Social, Mobile, Analytics and Cloud)(I-SMAC)*, pg. 741–749. IEEE, 2018.

Tian, Wei, Ruchi Choudhary, Godfried Augenbroe, and Sang Hoon Lee. "Importance analysis and meta-model construction with correlated variables in evaluation of thermal performance of campus buildings." *Building and Environment*, vol. 92, no. 1, pg. 61–74, 2015.

Tolan, Ghada M., and Omar S. Soliman. "An experimental study of classification algorithms for terrorism prediction." *International Journal of Knowledge Engineering-IACSIT*, vol. 1, no. 2, pg. 107–112, 2015.

Tsai, Chun-Wei, Chin-Feng Lai, Han-Chieh Chao, and Athanasios V. Vasilakos. "Big data analytics: a survey." *Journal of Big Data*, vol. 2, no. 1, pg. 1–32, 2015.

Van Der Maaten, Laurens, Eric Postma, and Jaap Van den Herik. "Dimensionality reduction: a comparative." *Journal of Machine Learning Research*, vol. 10, pg. 66–71, 2009.

Wu, F., L. Xu, X. Li, S. Kumari, M. Karuppiah, and M.S. Obaidat. "A lightweight and provably secure key agreement system for a smart grid with elliptic curve cryptography." *IEEE Systems Journal*, vol. 13, no. 3, pg. 2830–2838, 2018.

Xhemali, Daniela, Christopher J. Hinde, and Roger G. Stone. "Naïve bayes vs. decision trees vs. neural networks in the classification of training web pages." *International Journal of Computer Science Issues, IJCSI*, vol. 4, no. 1, pg. 6–23, 2009.

Xing, Eric P., Qirong Ho, Wei Dai, Jin Kyu Kim, Jinliang Wei, Seunghak Lee, Xun Zheng, Pengtao Xie, Abhimanu Kumar, and Yaoliang Yu. "Petuum: a new platform for distributed machine learning on big data." *IEEE Transactions on Big Data*, vol. 1, no. 2, pg. 49–67, 2015.

Xing, Eric P., Qirong Ho, Pengtao Xie, and Dai Wei. "Strategies and principles of distributed machine learning on big data." *Engineering*, vol. 2, pg. 179–195, 2016.

Zhou, Lina, Shimei Pan, Jianwu Wang, and Athanasios V. Vasilakos. "Machine learning on big data: opportunities and challenges." *Neurocomputing*, vol. 237, pg. 350–361, 2017.

Zhu, Rongchen, Xiaofeng Hu, Jiaqi Hou, and Xin Li. "Application of machine learning techniques for predicting the consequences of construction accidents in China." *Process Safety and Environmental Protection*, vol. 145, pg. 293–302, 2021.

7 Securing IoT through Blockchain in Big Data Environment

Muralidhar Kurni, Saritha K, and Mujeeb Shaik Mohammed

CONTENTS

DOI: 10.1201/9781003152392-7

7.1 INTRODUCTION

Technical specialists, digital administrators, marketing managers, writers, bloggers, and scholars have addressed and supported a new distributed paradigm for the safe processing and storage of transactions using blockchain technology. IDC FutureScape projected that blockchain would account for 20% of global trade finance by 2020 (Distributed Ledger Working Group, 2018). Coindesk estimates that venture capitalists have invested more than 1.8 billion dollars in blockchain startups (Distributed Ledger Working Group, 2018). Consortia and partnerships have formed, such as the Enterprise Ethereum Alliance, which define new applications across industries for blockchain technology.

Blockchain, public, and distributed ledger of transactions grouped into blocks promise to (Distributed Ledger Working Group, 2018):

1. Grow pace, efficiency, and security of digital asset ownership transfer
2. Eliminate the need to certify ownership and transparent transactions for central authorities
3. Reduce fraud and corruption with a straightforward and publicly auditable report
4. Reduce administrative costs through agreements to automatically trigger, safeguard, and certify trusted activities based on clear terms (smart contracts).

One primary challenge associated with blockchain adoption is identifying relevant uses to benefit from blockchain technology incorporation (Iansiti & Lakhani, 2017). The IoT (Internet of Things) has long been connected to security vulnerabilities and challenges, and experts and organizations have started to explore the use of blockchain for IoT security (Palo Alto, 2020). Organizations such as IOTA and the trusted IoT alliance have focused on IoT security blockchain (Trusted IoT Alliance, 2017).

IoT transforms customer behavior and market processes in its own right. Distributed IoT edge devices capture and distribute processing data (Ai et al., 2018). IoT systems use these data to provide advanced services, automation capabilities, and personalized user experiences. IoT networks are dynamic and distributed (Moeini et al., 2018). These include smartphones, mobile apps, gateways, cloud computing, analytics, machine learning (ML), networks, web services, storage, fog layers, and users. All these systems write and read data that can be documented on a ledger as transactions.

The Cloud Security Alliance (CSA) IoT Working Group (IoT WG) has been concentrating on recording best practices in IoT security since 2014 (Russell et al., 2015).

Due to the potential advantages of using blockchain technology to solve IoT security, the IoT WG has worked together with the CSA Blockchain/Distributed Ledger Technology Working Group to study and document some of the ways blockchain can contribute to securing IoT systems (Distributed Ledger Working Group, 2018).

- *Blockchain:* A technology facilitator that has powered progressive change and disruption through the digital economy, promoting quickly emerging cryptocurrencies such as BitCoin, Ethereum, Litecoin, and Dash. Blockchain's popularity as a foundation for cryptocurrencies has led to recent industry research into secure networks and applications using distributed ledger technology. Many business efforts in 2017 centered on the development of restricted prototypes and concept proof, most of which helped master the intricacies of this dynamic technology.
- *The IoT:* A rapidly maturing collection of technologies that support business and mission transformation processes. IoT's maturity has varied across and aided consumers, transport, energy, healthcare, manufacturing, retail, and finance. IoT consists of internetworking of physical devices such as industrial control systems, connected cars, systems for drones and robotics, smart buildings, and other elements integrated with sensors, actuators, electronics, software, and communication networks to facilitate data exchange between these items.

This chapter outlines an overview of high-level blockchain technology and sets out a series of architectural trends to render blockchain technology for IoT protection. Relevant use-case examples of the IoT protection blockchain have also been investigated, but these use cases' technological implementation will differ throughout businesses.

7.2 OVERVIEW OF BLOCKCHAIN TECHNOLOGY

Usually, this is a digital record or ledger. Any transaction in this record is authorized by the owner's advanced digital signature, which validates and prevents the transaction from being changed. Therefore, the value of the digital ledger is practically uncrackable (Namasudra et al., 2020). The digital ledger is like a Google spreadsheet maintained by multiple computers in a network, allowing transactions to be monitored. The main problem is that information can be copied, but it cannot be updated.

7.2.1 Why Is Blockchain Popular?

Assume you move money from your bank account to your family or friends. You will log in to online banking and pass the amount by using your account number to the other person. Transaction records are updated by the bank when the transaction is made. It looks easy enough, though. There is a potential problem that most of us overlook. Such transactions can be exploited very easily. Many who know this reality are always cautious about these kinds of transactions; thence, third-party payment applications are developed in recent years. However, this weakness is primarily due to the advancement of blockchain technology.

Technologically speaking, blockchain is a digital leader that has recently gained much recognition and momentum. However, why did it get so accepted? Okay, let us mine the whole idea into it. Data and transaction record keeping is a vital part of the business. This information is usually stored at home or distributed through a third party such as bankers, or brokers, henceforth increasing cost, time, or business. Prosperously, blockchain eliminates this lengthy process and enables quicker transactions, thus saving time and money. People mostly believe that blockchain and Bitcoin can be utilized mutually, but that is not necessarily the case. Blockchain is the technology that supports diverse applications in many industries, such as banking, manufacturing, and supply chain. Bitcoin is, however, money that's secure with blockchain innovation. In a progressively digital world, blockchain is an evolving technology with several advantages (Rejeb et al., 2019):

- *Highly secure:* A computerized digital signature to carry out fraud-free transactions makes it incomprehensible for others to corrupt or alter a person's data without a specific advanced digital signature.
- *Decentralized system:* You typically require the permission of administrative authorities such as a banking agency or government for transaction exchanges. However, blockchain transactions are made with mutual agreement among clients leading to safer, smoother, and faster transactions.
- *Automation capability:* It is programmed, and if the trigger conditions are met, it will automatically produce systematic actions, events, and payments.

7.2.2 How Does Blockchain Work?

Blockchain has three main concepts: blocks, nodes, and miners (Sultan et al., 2018).

1. *Blocks*: Each chain consists of several blocks, and each block consists of three essential elements:
    - The *data* in the block.
    - A whole 32-bit number called a *nonce*. When a block is formed, the nonce is generated randomly, and a block header hash is generated.
    - The *hash* is a 256-bit nonce number. It ought, to begin with, a vast number of zeroes (i.e., be minimal).
2. *Miners*: Miners are attacking the chain and then building a new block. There is a nonce value, and then an intermediate value is combined with the previous block's hash, which makes it hard to mine on large chains. Miners use special tools to solve a nonce's too complicated mathematical problem that generates the accepted hash. A nonce is 32 bits and a hash 256, so about four billion combinations of nonce and hash are required to verify the correct combination. If any block is modified earlier in the chain, the block must be reverified with the move. Blockchain technology makes it impossible to hack. Seeking golden nonces takes an immense amount of time and computing resources. If a block is validated, all network nodes support the move, and miners are paid.

7.2.3 Benefits of Blockchain Technology

Some of the benefits of blockchain technology are (Golosova & Romanovs, 2018):

- *Time-saving:* Central authority verification is not required for settlements making the process quicker and economical.
- *Cost-saving:* Blockchain network eliminates costs in many ways; intermediaries are limited as there is no need for third-party verification as members will share assets directly. Transaction efforts are lessened as every individual has a copy of the shared ledger.
- *Tighter security:* The system is protected from cybercrimes and fraud. No one can temper with blockchain data as millions of participants exchange it.

7.3 INTERNET OF THINGS

The IoT is a system of linked computing devices, digital machines, mechanical objects, animals, and persons with exclusive unique identifiers proficient in sharing data across a network without interactions between humans and humans or computers (Liau et al., 2006). Figure 7.1 shows an example of an IoT system. A *thing* on the Internet is some other natural object or human-made object with an IP address and the ability to transmit data through a network, like someone implanted with a heart monitor, an animal with a biochip transponder, and an automotive with built-in sensors to alert the driver for an abnormal condition (Hosain, 2018). Organizations in many

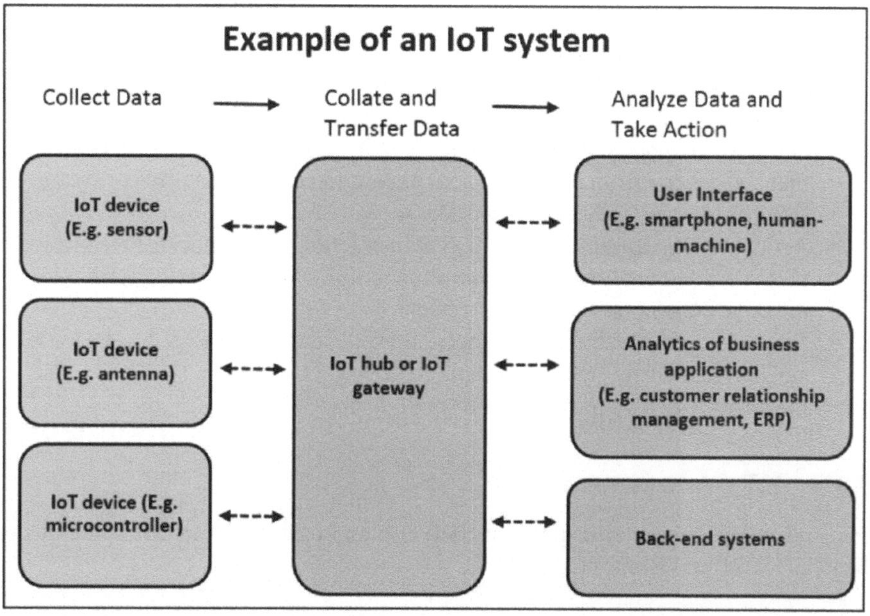

FIGURE 7.1 Example of an IoT system.

industries use IoT to run more efficiently, better understand their clients, implement better customer support, maximize market value, and enhance decision-making.

7.3.1 How IoT Works

An IoT environment contains web-enabled smart devices used to transmit, capture, and act on data they receive from their environment through embedded systems such as sensors, processors, and communications hardware. IoT devices share the sensor data they collect through an IoT gateway or distinct edge unit when the data is analyzed locally or sent to the cloud to analyze. Often, these devices communicate with similar devices and use each other's information. Without human involvement, these devices complete significant work despite people communicating with them—to access data and set up or provide directions (Ud Din et al., 2019). Network, organizing, connectivity, and communication conventions for these web-enabled devices rely significantly on the specific IoT applications. IoT makes Artificial Intelligence (AI) and ML more dynamic and simpler (Ghosh et al., 2018).

7.3.2 IoT—Key Features

IoT's main features include the following (Patel et al., 2016):

- *AI:* IoT makes almost everything "smart," which means that it improves every perspective of life by data collection capacity, network, and AI algorithms. It can be easy to upgrade your cabinets and fridge to detect when your cereal and milk go low and order your favorite foods.
- *Connectivity:* Latest networking technology, particularly IoT networking, suggests that networks are no longer mainly linked to significant providers. Networks can be much cheaper and smaller although being still practical. IoT between its system devices creates these small networks.
- *Sensors:* Without sensors, IoT fails its perfection. They describe devices that convert IoT from being a typical passive interface into an active system that can interact with the real world.
- *Active participation:* Much of current interaction with applicable technologies is carried out by passive commitment. IoT provides a new criterion for effective content or service engagement.
- *Small devices:* Devices, as predicted, have, over time, emerged to become smaller, less costly, and influential. To deliver its precision, scalability, and flexibility, IoT uses purpose-built small devices.

7.3.3 IoT—Advantages

IoT's advantages cover all aspects of daily life and business. Here are some of the benefits IoT offers (Brous et al., 2020).

- *Improved customer engagement:* Current analytics endure from blind dazzle spots and major accuracy failures, and engagement remains inactive. IoT transforms this to be more effective and makes engagement with audiences richer.

- *Technology optimization:* The innovations and data that enhance consumer experience improve the system to use and lead to more potent technology improvement. IoT launches vital imperative, functional, and field informative data environments.
- *Reduced waste:* IoT specifically defines regions of change. Current analytics gives us superficial insight, but IoT provides real-world knowledge, which leads to better resource administrative management.
- *Enhanced data collection:* Modern data collection has drawbacks in its architecture for submissive use. IoT breaks it from those positions and takes it where people want to analyze our world. It gives a precise representation of everything.

7.3.4 IoT—Disadvantages

Although IoT provides an incredible variety of advantages, it also poses several challenges. Some biggest problems are listed here (Tawalbeh et al., 2020).

- *Security:* IoT establishes a persistently connected devices' ecosystem that interacts with the networks. Despite any security measures, the system provides little control. This helps users to be exposed to various types of attackers.
- *Privacy:* Sophistication of IoT offers considerable personal data in intense detail without the user's active involvement.
- *Complexity:* A few IoT systems are complicated by using various technologies, a wide variety of emerging technology in design, implementation, and maintenance.
- *Flexibility:* The IoT framework's flexibility to assimilate easily with one another is a problem for many. They are concerned that they have many conflicting or locked structures.
- *Compliance:* IoT must comply with regulations like any other technology in the business sector. Its sophistication renders enforcement unbelievably difficult as many regard regular compliance with software as a battle.

7.4 THE IOT SECURITY CHALLENGE

Although security issues are not contemporary in information technology, fresh and unique security problems arise due to many IoT implementations' characteristics (Jurcut et al., 2020). A fundamental focus must be to resolve these problems and ensure protection in IoT products and services. Users must be assured that IoT devices and associated data services are safe from vulnerabilities, particularly as this technology is becoming prevalent and incorporated into our everyday activities. Weak-protected IoT devices and services will serve as possible input for cyberattacks and expose user data to theft by insufficiently securing data streams (Fawaz & Shin, 2019).

IoT devices' interconnected existence means that any poorly protected online device will globally affect Internet security and resilience (Rainie & Anderson, 2017). This issue is compounded by factors such as the mass deployment of homogeneous

IoT devices, specific devices' ability to connect automatically to other devices, and the possibility of these devices being used in a hazardous environment. It is up to all the IoT developers and consumers to ensure their devices and systems do not compromise the Internet. There needs to be a collaborative approach among security experts, organizations, and researchers to ensure IoT protection.

7.4.1 A Spectrum of Security Considerations

It is important to note that protection is not absolute when considering IoT devices. This device protection is not a stable or vulnerable binary proposition. Instead, IoT security as a continuum of system insecurity can be conceptualized.

The scope ranges from entirely insecure devices without security features to highly secure systems with several safety layers. New security threats are emerging in a limitless game between cat and mouse, with computer manufacturers and network operators continually adapting to new threats. The IoT's overall protection and durability depend on manipulating and measuring security risks (Butun et al., 2020). Security is based on the probability of compromise, the harm from a compromise, and the time and money needed to fulfill such security criteria. If an individual cannot tolerate a high degree of risk of injury, the person will feel justified in spending a large amount of money protecting the system or device from attack. If a more complicated security model is also more costly or difficult to maintain, she might not be able to afford it.

Several factors influence the assessment and mitigation. To decide whether the safety issue outweighs the benefit, consider the damage, the cost to fix the harm, and the possible benefits of addressing the concern (Rose et al., 2015). An enterprise often implements combinations of uninterruptible power supply and high availability systems. The IoT devices' networked connectivity ensures that locally taken security decisions can affect other devices. In theory, creators of intelligent objects for the IoT are obligated to ensure that such devices do not place their users or others at risk. Company and business are concerned with reducing the cost, complexity, and time to market of vendors. For example, high-volume, low-margin IoT devices that already have additional costs for goods into which they are incorporated become very frequent; the addition of more memory and a faster processor to implement security might easily make this product commercially uncompetitive.

Economically speaking, the lack of IoT devices' security contributes to a negative externality, where costs are levied on other parties by one party or parties. A typical example is an environmental pollution, where all parties are responsible for the environmental degradation and clean-up costs (negative externalities) of polluter actions. The problem is that the effect of the external charges levied on others is generally not considered in the decision-making process unless, as with waste, the polluter is imposed a tax to force him to minimize pollution. As Bruce Schneier has discussed, for information protection (Rose et al., 2015), externality exists when the vendor supplying the product does not bear the vulnerability cost. Liability law can, in that case, affect the externality of vendors and produce additional safety products. These safety issues are not new in information technology, but they are essential because of the scale of specific challenges emerging in IoT implementation, as discussed in the following sections.

7.4.2 Unique Security Challenges of IoT Devices

IoT devices tend to vary in meaningful ways that call for protection from conventional computers and computing devices (Rose et al., 2015):

- Many Internet-connected devices, such as sensors and consumer products, are installed beyond traditional communication devices. With the paired devices, the connections with additional devices are unprecedented. Many of these devices may also connect and to devices without permission. Therefore, new considerations can be required for existing resources, methods, and strategies associated with IoT protection.
- Most IoT implementations are similar or nearly identical. It makes it easy for hackers to exploit a single vulnerability by using many similar devices. Vulnerabilities can apply to all products that use a single communication protocol or all products using a shared communication protocol.
- Many IoT devices can be used for many years longer than average for high-tech equipment with anticipated service life. Furthermore, in situations where reconfiguration or updating is difficult or impossible, these devices can be deployed, or they may survive the business that created them, leaving orphaned devices free of long-lasting support. These scenarios demonstrate that appropriate protection measures for deployment will not be sufficient for the system's entire lifetime as security threats emerge. As such, vulnerabilities will continue for a long time. This is contrary to the conventional computer systems model that is usually upgraded to resolve security threats with operating system software upgrades over the entire computer life. Long-term maintenance and IoT system management are a big security challenge.
- Many IoT devices are deliberately designed without simple upgrading or a complicated update process. Companies, such as Fiat Chrysler, are now giving recalls of a vehicle (being linked wirelessly) due to weakness. The vehicle's firmware must be modified manually, or the user must upgrade himself with a USB key. Moreover, a high percentage of cars will not be modified because of the inconvenience for owners and will be permanently exposed to cybersecurity risks, mainly when the car appears to work differently.
- Many IoT devices are hard to see because they are small and have many data flowing through them. It poses a security risk if IoT devices perform functions unintended by the user or collect more data than the user wishes. When the system is modified, any improvements made to the manufacturer's functionality can be available to the user.
- Some IoT devices can be installed where there is difficult or impossible physical protection. Attackers can access IoT devices directly physically. To ensure protection, antitampering features and other design advances must be taken into account.
- Some IoT devices are designed to be unobtrusively incorporated into the environment, like many environmental sensors, where a user does not

consciously recognize or control the device's operating state. Furthermore, devices cannot explicitly warn the user when a security issue occurs, making it impossible for a user to realize when an IoT device has been broken. If correction or mitigation is even necessary or feasible, a safety violation can continue for a long time until it is detected and rectified. Similarly, the user might not be aware of a sensor's presence in its environment, allowing a security violation to proceed for long periods without detection.

- Early IoT models assume that IoT is a result of large private and/or public technology companies, but "Build Your Own Internet of Things" could become more popular, as illustrated by the increasing communities of developers Arduino Raspberry Pi60.

7.4.3 IoT Security Questions

There has been a range of concerns about security issues raised by IoT devices. Many of these questions existed before the IoT development, but they increased in importance because of IoT devices' size. Such notable issues are (Rose et al., 2015):

- *Acceptable practices in design:*
  - What are the best practices for engineers and developers to build IoT devices to make them secure?
  - How are security concerns from IoT devices captured and transmitted to emerging communities to ensure the next generation is even more secure?
  - Which training and educational tools are available for teaching engineers and developers to develop more secure IoT designs?
- *Cost versus security trade-offs:*
  - How do stakeholders make informed decisions on cost–benefit analysis for IoT devices?
  - How do we measure and analyze security risks accurately?
- *Metrics and standards:*
  - How do we effectively define and calculate IoT system security characteristics?
  - How do we assess the efficacy of security and countermeasures of the IoT?
  - How will the best security practices be implemented?
- *Data confidentiality, authentication, and access control data:*
  - What is the optimal function of IoT system data encryption?
  - Is it necessary to use strong IoT device encryption, authentication, and access control technologies to avoid eavesdropping and detecting data stream attacks generated by these devices?
  - What encryption and authentication technologies could be used for the IoT, and how could cost, size, and processing speed be applied in the sense of an IoT device's limitations?

    - What are the foreseeable management problems that IoT-scale cryptography needs to address?
    - Are there problems with crypto-key lifecycle management and the planned time during which any algorithm should remain secure?
    - Are end-to-end processes secure enough to be used by traditional consumers?

- *Field upgrade capacity:*
    - Should devices be configured for maintenance and upgradeability in the field to respond to emerging safety threats with an extended lifespan required of several IoT devices?
    - A centralized security management system could install new software and parameter settings on a fielded IoT device if each device had an integrated device management agent. However, management systems add costs and complexity; could other software-updating approaches be consistent with IoT devices' extensive use?
    - Are there groups of low-risk IoT devices that do not guarantee this type of feature?
    - Are IoT devices exposing user interfaces (usually purposely minimal) adequately controlled for system management considerations (by anyone, including the user)?

- *Shared responsibility:*
    - How can sharing mutual responsibility and cooperation for IoT security between stakeholders be encouraged?

- *Regulation:*
    - Should manufacturers of products be penalized with known or unknown security bugs for selling software or hardware?
    - How can product liability and consumer protection regulations be tailored to cover all negative externalities about the IoT and function in a cross-border environment?
    - Will legislation keep pace and be successful, given the changing IoT technology and safety threats?
    - How can laws be balanced against the needs of creativity, freedom of the Internet, and freedom of speech without permission?

- *Device obsolescence:*
    - What is the best solution as the Internet grows and security threats shift with outdated IoT devices?
    - Should IoT devices have an integrated end-of-life expiry feature to uninstall them?
    - Such a requirement could result in older noninteroperable devices being forced out of operation in the future and replacing them with safer and more interoperable devices. This will be very difficult in the free market. What are the effects of automated IoT decommissioning devices?

The list of questions includes general security problems for IoT devices. However, it must be noted that when a machine is on the Internet, it is also part of the Internet, which means that only if the users of such devices apply an effective mutual security strategy, acceptable solutions are obtained.

The collaborative model has emerged to provide a practical approach to secure the Internet and cyberspace, including the IoT, among industries, governments, and authorities.

7.5 IS BLOCKCHAIN THE SOLUTION TO IOT SECURITY?

IoT devices' security is a continuing concern. It leaves open an expansive room for hacking devices. This kind of security dilemma is a big one for smart homes and smart vehicles (Allhoff & Henschke, 2018). An IoT device could be hacked, for example, by a hacker who could take over a self-driving vehicle. Strong protection is a must for all data transmitted from IoT devices.

Although there are various IoT security guidelines, including biometrics and two-factor permission, blockchain IoT security is one possible solution (Atlam & Wills, 2020). For Bitcoin and Ethereum, blockchain provides an attractive IoT security solution. The blockchain provides strong data security, locking access to IoT devices and allowing a shutdown of infected devices in an IoT network (Alotaibi, 2019). Hyundai has recently funded a blockchain start-up developed for IoT security. This creative strategy, named HDAC (Hyundai Digital Access Currency), establishes a private network permitted.

Thomas Hardjono, MIT Connection Science Chief Technology Officer, indicated that infrastructure must handle devices and data access. In a recent paper, he proposes a ChainAnchor blockchain-based IoT system. The framework proposes access layers that can maintain nonenabled devices or cut out bad actors (such as hacked devices) from the network. It also provides cases for securely selling and removing devices from a blockchain. This framework deals with system protection with device manufacturers, service providers, and independent third parties (Compton, 2017).

However, if IoT security is to be blockchain, there are problems to solve. For one thing, blockchain mining requires a great deal of computing power. Many IoT devices do not have the necessary power. Current blockchains are vulnerable if a group of miners dominates more than 50% of the network's hash rate. This is very difficult since nodes are spread globally in a traditional blockchain. However, the computing capacity of a home IoT blockchain could be more easily hacked. IoT safety will continue to grow as regulations on their growth continue and their advancement continues. However, the prospect of a blockchain IoT protection framework has a great potential.

7.6 WHEN THE IOT MEETS BLOCKCHAIN

The IoT has proliferated over the years and has connected multiple devices and networks to the Internet (Khanna & Kaur, 2020). On the other hand, the decade-old blockchain is intended to revolutionize business models with its encrypted and distributed ledger designed to create manageable and real-time records (Zhang et al., 2019). The blockchain aims to store the verifiable and secure records of all IoT devices and processes.

Blockchain functions as a distributed ledger in which any deletion or alteration of data is registered and a long chain of events is generated as more entries, namely, blocks. Each transaction is followed and can never be updated or removed by a digital signature. Given its decentral existence, blockchain can prevent a vulnerable device, either an intelligent home or a smart factory, from pushing false information and disrupting the networking setting.

As part of recent notable IoT safety accidents, blockchain will mitigate DDoS attacks' risk affecting several devices once the device's failure does not impact other devices (Salim et al., 2020). Such convenience is crucial for preserving connectivity and functionality in facilities, particularly critical networks, in securing smart cities. With blockchain, each device has strong encryption, guarantees safe communication with other devices, and provides anonymity when IoT is most concerned (Atlam et al., 2020). Adopters will be able to monitor devices better and deliver security updates, helping to improve insecure devices.

The combination of blockchain and IoT could also resolve supervision (Banerjee et al., 2018). For example, transactions from various sources can be handled in companies using a clear and permanent record, in which data and physical products are monitored through the supply chain. If an incorrect decision or machine overload happens, the blockchain record can define the point—say, a device or a sensor—where anything went wrong, and the organization will respond instantly. Blockchain can also reduce operating costs as it removes intermediaries or middlemen (Witzig & Salomon, 2019).

7.7 IOT ARCHITECTURAL PATTERN BASED ON THE BLOCKCHAIN SERVICE

Adapting blockchain technology to the IoT requires the study of a blockchain-based IoT architectural pattern. As mentioned in the following sections, the specified pattern should include three components (Distributed Ledger Working Group, 2018):

7.7.1 Model of Communication

The communication model defines the installation of the applications directly into IoT nodes and/or in the cloud with APIs to the IoT nodes. Figure 7.2 shows the popular and agreed blockchain technology and IoT model, when the IoT edge systems have robust capabilities, allowing them to host the transaction node's software, store the ledger, and maintain network-wide communication.

IoT transaction nodes: In Figure 7.1, each IoT system hosts the leader and can participate, including mining, in blockchain transactions. Each device has a private key or includes functions for the internal generation of a private key for network transactions. This end-state model offers three essential capabilities that a blockchain service enables:

- The network of autonomous IoT devices' (e.g., consensus and peer-to-peer messaging) autonomous coordination.

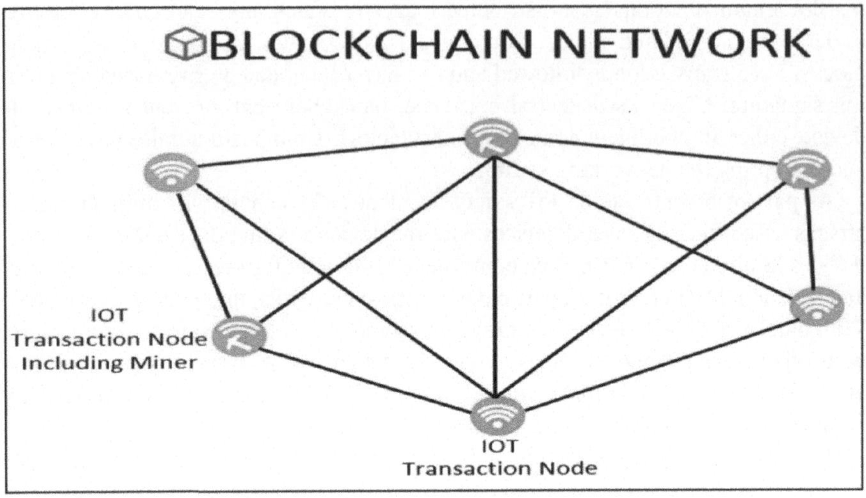

FIGURE 7.2 Each IoT node acts as a blockchain transaction node.

Source: Distributed Ledger Working Group (2018)

- A transaction leader where any IoT computer can build a cryptographic transaction.
- A distributed database with a modified version of the ledger on every IoT device.

Cloud-enabled IoT blockchain network: Transactions and mining nodes are placed in the cloud and on-site in a cloud-enabled blockchain network. The nodes can be corporate servers, company/personal computers, smart devices (e.g., phones or tablets), cloud-based virtual machines, and IoT devices with enough hardware (CPU, RAM, storage, etc.). Figure 7.3 shows an overview of cloud-enabled IoT blockchain network.

IoT devices with limited hardware resources serve as blockchain clients. You should not store the booklets. These clients interact with the blockchain via API. APIs can be HTTP, REST, or JSON.

IoT devices collect data relating to the blockchain service transaction nodes or engage in smart contract transactions by pointing to cloud-based blockchain nodes. IoT devices are still supplied with private keys to sign their data in this way. The signed data was forwarded upstream to the processing nodes. You must have a different trust relationship between the IoT system and the transaction node. One system may be whitelisted, while another is secured by a two-way authentication. Hardware encryption keys can be used to store private keys securely.

Access to mining nodes can be limited to designated operators for a licensed area (private blockchain service). Participants may decide to adopt this architecture pattern to enhance protection or enforcement purposes in a consortium area (partially authorized blockchain service) or a permitted area (private blockchain service).

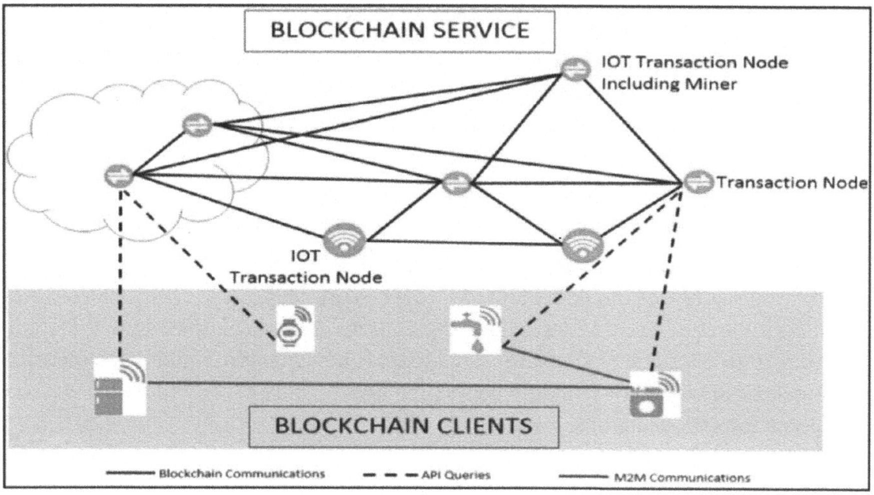

FIGURE 7.3 Cloud-enabled IoT blockchain network.

Source: Distributed Ledger Working Group (2018)

Bitcoin implementation provides this kind of feature with "thin clients," also known as simplified payment verification that does not store a full copy of each ledger block. Those "thin clients" use a Bitcoin Client API to communicate with a server.

Messages between multiple IoT devices can be exchanged. These messages contain data incorporated into IoT transactions participating in the exchange of transaction nodes. Communication protocols and message formats between IoT devices are outside the blockchain implementation reach: these communications apply to machine-to-machine communication such as MQTT (Message Queue Telemetry Transport).

7.7.2 A Rich Ecosystem for Leveraging Interoperability Capabilities

It will be an opportunity to speed up its adoption by creating an ecosystem around blockchain technology. This is what we are talking about. The potential ecosystem will offer IoT integration in blockchain service streamlined capabilities.

- *Service providers*, such as Blockcypher, have API capabilities that simplify IoT/blockchain customer/service interactions. API intermediation enables creating IoT features that connect with various blockchain services based on service value instead of technical implementation of blockchain technology.
- *Solution providers*, including Credits, offer frameworks for creating a private blockchain service quickly. These systems are based on nodes of transactions. Clients have access to each node through APIs. These systems can also communicate with other blockchain services.

7.7.3 Cohabitation between Multiple Blockchain Services

As seen in Figure 7.4, a different definition focuses on several blockchain platforms providing different characteristics and currencies and will improve. These blockchain platforms provide additional features. Each blockchain service can be natively attached to others or use third-party APIs.

7.7.3.1 The IoT Architecture Pattern Based on Blockchain Technology

The CSA IoT and Blockchain/Distributed Ledger Technology Working Groups propose the following system (Distributed Ledger Working Group, 2018) under which IoT clients operate in multi-blockchain applications. Figure 7.5 shows the IoT architecture pattern based on blockchain technology. A new industry initiative to incorporate blockchain technology mainly adopts the architecture under which IoT devices are clients for a blockchain service.

7.8 FEATURES TO CONSIDER WHEN SECURING THE IOT USING BLOCKCHAIN TECHNOLOGY

Blockchain can help secure IoT devices. IoT devices can be optimized to use public blockchain services or communicate over a protected API with private blockchain nodes in the cloud. Integrated blockchain technology in an IoT system's security framework enables IoT devices to safely reveal each other; encrypt machine-to-machine transactions with distributed vital management techniques; and verify software mage legitimacy, authenticity, and policy updates.

Based on the possible architectural patterns detailed in this chapter, an IoT system interacts with a blockchain transaction node via an API, enabling even restricted

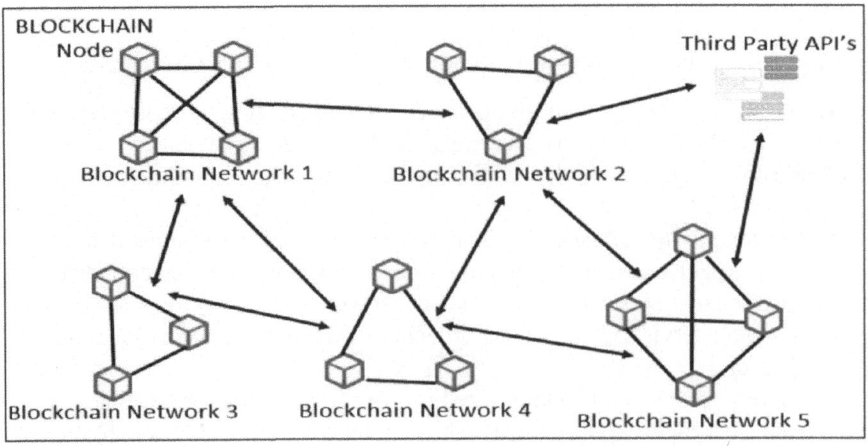

FIGURE 7.4 Each blockchain service can run in different contexts such as personal home networks, enterprise, and the Internet.

Source: Distributed Ledger Working Group (2018)

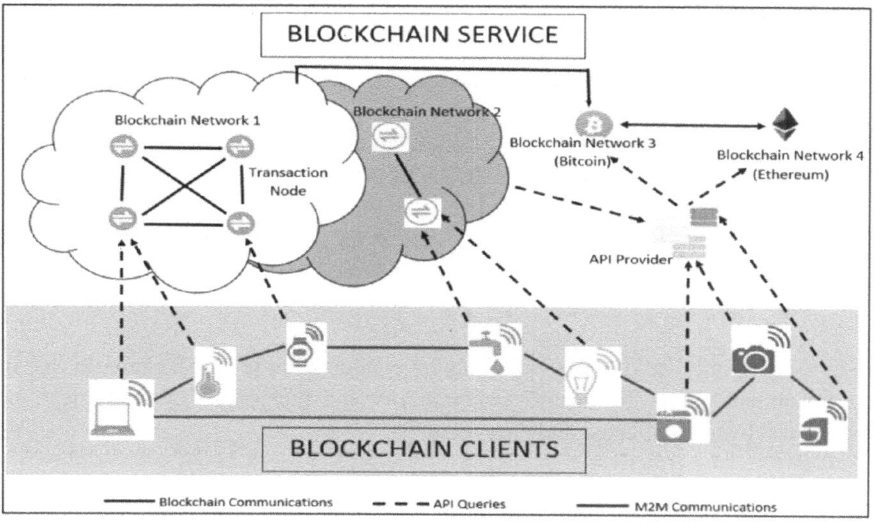

FIGURE 7.5 The IoT architecture pattern based on blockchain technology.

Source: Distributed Ledger Working Group (2018)

devices to participate in a blockchain operation. To ensure security, attention should be focussed on a specific blockchain service during the bootstrapping of an IoT device.

Following is an IoT discovery use case that enables an IoT device to be registered in a transaction node. First, the IoT system must have credentials that show authorization to be added to a transaction node. This credential must be given in a secured environment that protects against the threats of a specific ecosystem of IoT devices. Blockchain technology and available business initiatives show five features to resolve when securing IoT using blockchain technology (Distributed Ledger Working Group, 2018).

7.8.1 Scalable IoT Discovery

Smart cities and large-scale IoT implementations will result in thousands or tens of thousands of IoT devices enabled jointly. In autonomous machine-to-machine transactions, these devices also communicate with each other. The devices will need to be able to find legitimate peers and interacting services. IoT systems may benefit from the use of public and private blockchain implementations in scalable IoT discovery.

For example, with Bitcoin, a collection of hard-coded DNS seeds offers bootstrap services to new users and computers. The DNS seeds can be preconfigured on an IoT system. IoT devices query these addresses, and the full node IP address is given. The IoT device then logs into a node and calls for a list of additional IoT devices on the network. When provided, the IoT system will initiate peer-to-peer communication while promulgating peer exploration information within the network. The presetting of the DNS seed addresses reduces the ability to execute a man-in-the-middle attack. Until selecting a node, IoT devices obtain information from several DNS seeds. DNSSec must be used to protect root server name resolution and to mitigate attacks by DNS spoofing. DNS seed addresses should be hard coded in the firmware.

A private blockchain service can also support IoT system bootstrap and network enrollment. Transaction nodes authenticate IoT devices until a trustworthy node list is given. IoT systems are issued with the following entry credentials:

- Security credentials installed or created internally in the IoT device must be generated and made accessible through a protected process that may be part of blockchain implementation.
- The credentials issued by the IoT device owner or installation technician initialize the device's registration on a secure server to access such IoT credentials.

In any event, only valid IoT devices may be connected to the blockchain service in the enrolment phase. To ensure confidentiality and integrity, all communications listed must be authenticated and encrypted.

7.8.2 Trusted Communication

In some cases, for instance, public use, IoT devices require the use of a protected communication channel to share the data necessary to construct the transaction to be stored in the ledger. You may also use this ledger to store public encryption keys. If exchanges need to be maintained confidentially, an IoT device (sender) transmits an encrypted message to a peer IoT (receiver) device using the recipient IoT device's public key, stored in a blockchain service. The IoT sender requests its transaction node to move from the blockchain ledger to the IoT receiver's public key to facilitate this secured transaction. IoT sender will encrypt the message with the IoT recipient's public key. Figures 7.6 (a) and 7.6 (b) show the phases involved in the trusted communication.

The message can only be decrypted by the recipient using his private key. Key agreements algorithms such as Elliptic Curve Diffie Hellman must be used for generating transaction authentication keys such as content encryption keys (CEKs) and traffic encryption keys (TEKs) to secure transactions.

In this case, the blockchain service functions as an infrastructure of the distributed public key. Public keys will be stored during transactions: when a new IoT device is enrolled in a private or public service, a new transaction will be created. This exchange consists of IoT properties and their public key. Reregistration happens if an IoT device needs to renew its license. Revoked certificates may also be applied as transactions to the blockchain service. The protected transaction history in the directory provides the continuity of IoT device keys over their lives.

In a blockchain implementation, there are many types of cryptographic keys used. Wallet keys are also known as keys to secure blockchain transactions. The keys discussed in this use case represent identify keys and generate CEKs or TEKs.

- *IoT identity keys:* Asymmetric key pair used to establish key content-encryption material and traffic flow between IoT devices
- *Wallet keys:* Used to protect transactions in the ledger, IoT identification keys can include

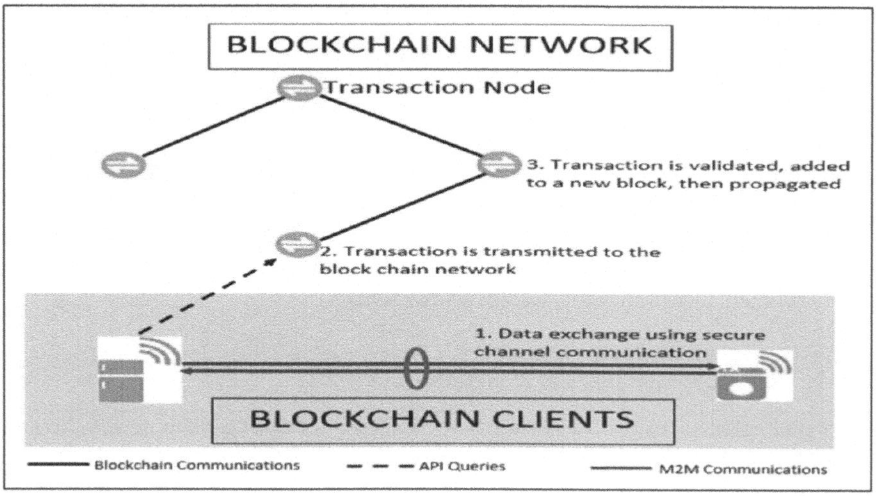

FIGURE 7.6(A) Trusted communication.

Source: Distributed Ledger Working Group (2018)

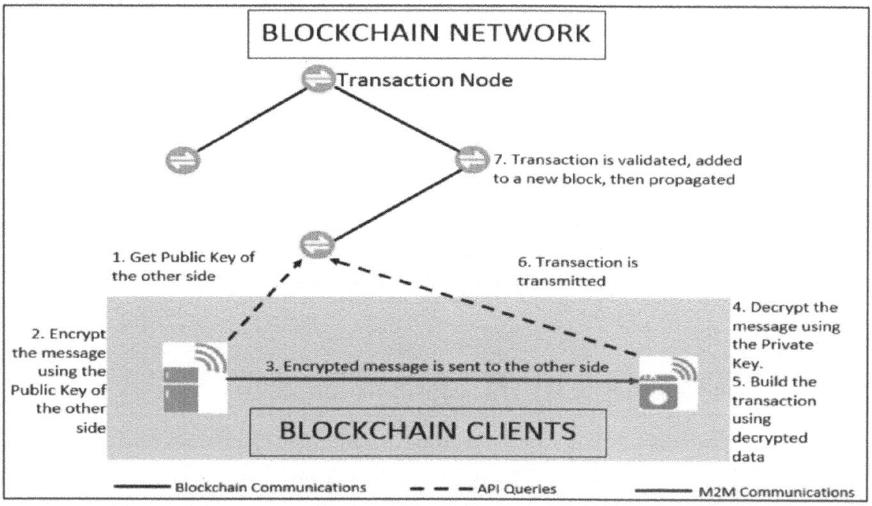

FIGURE 7.6(B) Trusted communication.

Source: Distributed Ledger Working Group (2018)

7.8.3 Semiautonomous Machine-to-Machine Operations

The machines' ability to work together in semiautonomous ways to accomplish a specific purpose is a core IoT technology feature. Blockchain can serve as a security facilitator for these independent transactions with smart contract features. Smart contracts can be written to include rules, fines, and contract conditions. Edge IoT devices can then be set up with an API to communicate with an intelligent contract to conclude agreements with peer devices and/or services. Each transaction must comply with the contract terms before execution, and all transactions must be written to the blockchain.

Intelligent contracts can impose restrictions on who can enter into transactions (which IoT devices). With the IoT node's wallet key, each transaction must be signed and wallets stored in hardware security containers. Blockchain transaction recording guarantees that transactions will not later be repudiated (e.g., when the IoT service provider enters a customer IoT system transaction).

7.8.4 IoT Configuration and Updates Controls

In trusted, stable settings, blockchain's technology is exciting, as more IoT devices are natively connected to cloud services. Following are three approaches to security:

1. IoT properties such as the current version of validated firmware and configuration information can be found in the ledger. The IoT system asks the transaction node to get its setup from the ledger during bootstrap. The configuration should be encrypted in the ledger to prevent the IoT network's topology by analyzing the public ledger's content.
2. The ledger will host the hash value for each IoT device of the latest configuration file. The IoT system downloads a trusted cloud service configuration file every night for some period and then uses the API transaction node to locate and match the hash value stored on the blockchain. This enables administrators to periodically flush wrong configurations and reboot new IoT devices on their network.
3. As described in point 2, the same process can be used for IoT device firmware images although additional bandwidth capacity may be needed at the point on the IoT device.

7.8.5 Stable Firmware Image Distribution and Upgrade

Blockchain technology will also support the trusted imaging process for IoT devices, similar to supporting installing familiar configurations from the cloud service provider. An IoT device developer can also implement its blockchain through IoT device firmware or use a public blockchain. The developer will hack the current known trusted images and load these hashes into the blockchain for its device families. This approach supports the improved protection of IoT devices in three ways:

1. You can customize IoT devices via the API to periodically download new firmware images. As most IoT devices do not have to store or retain data in memory, they can be overwritten if appropriate. For example, a regular or weekly image update could be allowed by validating the image hash against the vendor's blockchain.

2. To verify all vendor updates, IoT devices may use a blockchain image update mechanism.
3. IoT devices can either use method 1 or 2 to validate all updates and require the system owner's firmware update approval (by a safe method).

IoT manufacturers should strengthen the traditional approaches to software signatures by saving the firmware's digital signature to the directory rather than posting it on their website. Before the update is applied, IoT devices obtain the latest firmware's digital signature from the ledger and then validate it with a public maintenance key. This public maintenance key could be fused to the fabric/hardware level (no change/update features). A manufacturer's privately owned key must be guarded to ensure that all firmware is not compromised. A private-key attacker might make malicious firmware with an obviously "valid" digital signature visible. The process of modifying the manufacturer's public key on all devices takes considerable effort.

7.8.5.1 Firmware Reputation-Based Upgrade (Chain of Things)

A ledger transaction history capabilities will incorporate new firmware through a group of experts to strengthen confidence in avoiding the installation of malware corrupted firmware and backdoor firmware. The IoT devices' owner/administrator can also configure an automated IoT update when the firmware's credibility is at a particular stage in the ledger. This "acceptation" of the IoT device in the blockchain service may be based on the firmware credibility of the device in the ledger and would benefit from the following points:

- Avoid linking insecure devices to a blockchain service
- Implement IoT protection upgrade processes
- Set minimum security standards for blockchain service

Figure 7.7 shows the summary of blockchain security services for the IoT.

7.9 VARIOUS WAYS TO STRENGTHEN IOT SECURITY WITH BLOCKCHAIN TECHNOLOGY

The following points should be considered for a stable application of IoT (Sreevani, 2019).

1. *Secure communication:* IoT devices must communicate and store data directly to share data necessary to process a transaction. There are several ways to store encryption keys. The IoT framework sends a public key of the destination device stored in the network blockchain. The sender requests his node to retrieve the recipient's public key and then encrypts the message using the recipient's public key, allowing only the recipient to decrypt the message.
2. *User authentication:* Before sending them to other devices, the sender digitally signs the message. The receiving system uses a public key from the

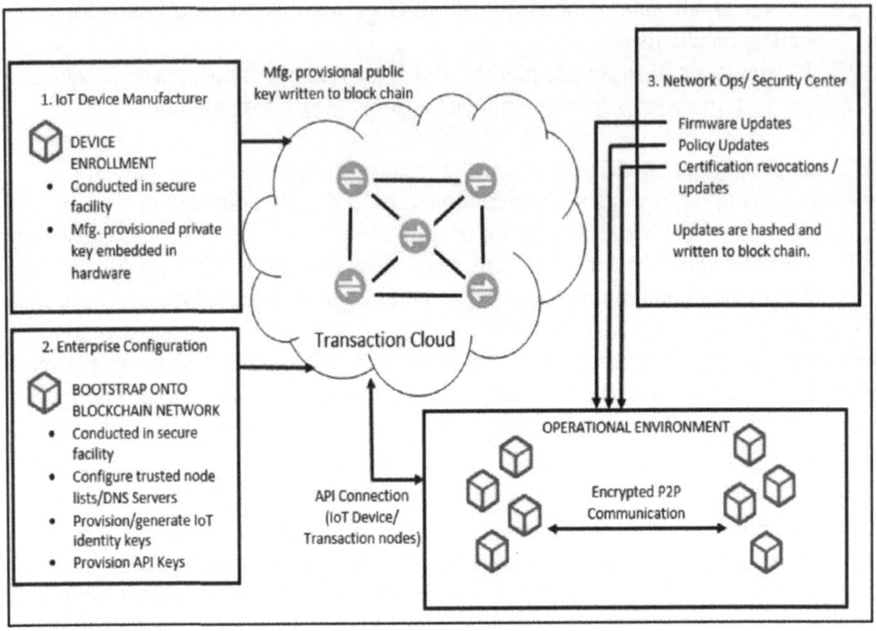

FIGURE 7.7 Summary of blockchain security services for the IoT.

Source: Distributed Ledger Working Group (2018)

directory and validates the received message's digital signature. The following is the digital signature work:

- The sender calculates the hash of a message, which then has a private key encrypted.
- The digital signature is transmitted along with the post.

3. *Discovering legitimate IoT at a large scale:* When a new IoT system is launched, root servers are asked to list confident network nodes. This device then registers in a node, and the information exchange begins. DNSSec needs to be introduced to protect the root server name resolution by preventing spoofing attacks. Each contact must be authenticated and efficiently encrypted, which can be achieved on the basis of the following points:

   - During installation, credentials are already installed on the device.
   - The IoT device owner can provide credentials.

4. *Configuring IoT:* Blockchain technology is a massive aid in creating a trustworthy and stable IoT system setup. Approaches that tend to be relevant here are given as follows:

   - The current ledger will host the hash value of the new configuration file. The IoT device must regularly upload a new trusted configuration file using a cloud service. The device then uses the blockchain API node to match the hash value in the blockchain. This allows administrators to regularly reboot IoT systems for the most recent trustworthy configuration on the network.

7.10 CHALLENGES IN INTEGRATING BLOCKCHAIN INTO THE IOT

Blockchain technology has made strides into the IoT, but it faces several obstacles (Trend Micro, 2018). The basic principle of a blockchain is the sequence of transactions. The chain is created by keeping a reference to past exchanges. That said, blocks can be challenging to build due to their size and complexity. It is more difficult to rearrange single blocks: Altering it with the previous block and changing the chain.

This setup seems to secure the IoT. Nevertheless, the IoT devices are moderately underpowered, and the blockchain protocols produce overhead traffic, creating a block-introducing latency. A problem exists, particularly for resource-constrained and bandwidth-limited systems, which need real-time updates or speedy responses.

Researchers have studied risks related to privacy, authentication, accessibility, and access control.

Also, the threat actors will exploit blockchain to expand their strategies of evasion. IoT sensors and devices may be exploited to relay incorrect data to a blockchain. The technology will be registered in the chain if the entry is validated. Adopters must also ensure that sensors and devices are ready to override in a compromise and allow only users who need to be regulated to log in.

7.11 SECURITY RECOMMENDATIONS

If operated correctly, blockchain can benefit IoT systems greatly by reducing costs and increasing performance. The penetration of technology into IoT-enabled environments is even far from being optimum. For example, IoT sensors are expected to integrate up to 10% of output blockchain ledgers by 2020. Also, many IoT systems will need to be computationally competent enough to handle voluminous blockchain implementations.

Although the single failure point has not been removed, the IoT protection endures continuous security implementation for all connected devices. In addition to the up-to-date software updates to avoid downtimes, organizations, IoT administrators, individuals, and like can look at multilayered security from the entrance to the end, with end-to-end protection, which avoids possible intrusions and compromises in the network. This includes (Trend Micro, 2018):

- *Changing default credentials:* The standardized credentials in factories have famously allowed IoT botnets to compromise connected devices. Users are recommended for using password protection to reduce the possibility of computer hacking by using unique and complex passwords.
- *Strengthening router protection:* A liable router renders a vulnerable network. Routers with robust security explications make it easier for users to pile up all associated devices while protecting privacy and productivity.

- *Security system setup:* Default device settings should be tested and updated to the user requirements. For improved security, needless customization and disabling features are recommended.
- *Network traffic monitoring:* An aggressive network traffic scanning will help users prevent malicious attempts. Security solutions in real-time scanning can also use automatic and effective malware detection.
- *Implementing added security measures:* Users are advised to allow firewalls and practice the additional safety protocol Wi-Fi-protected Access II Web credibility solutions, and application management also has greater network visibility.

7.12 USE CASES OF BLOCKCHAIN MECHANISMS FOR IOT SECURITY

This section presents various examples of blockchain mechanisms for IoT security (Khvoynitskaya, 2020).

The application of blockchain into IoT protection enables the direct exchange of information within connected devices rather than central network communication, thus reducing IoT's vulnerability to cyber securities. According to a Gartner report, the most current blockchain adoption rates among IoT companies in the United States are pharmaceutical, transport, and energy sectors. All these enterprises rely upon the transporting of physical goods, and the most popular blockchain associations are IoT veterans.

The information is screened directly on the device by a private key and anchored in a public blockchain, thereby storing data on each connection to a specific sensor forever on a ledger. The implementation of smart contracts also opens up new ways to strengthen this.

Such IoT implementations can also substantially raise standards in distinct sectors, such as the food and pharmaceutical industries. Several medical devices, for example, must be kept under regulatory temperatures. For several years, IBM's food trust has been on the job, but IoT sensors, in conjunction with blockchain, will provide even more exciting opportunities.

IBM and Samsung have developed a PoC for ADEPT (Autonomous Decentralized Peer-to-Peer Telemetry) for their blockchain-enabled IoT device. It uses smart contracts and peer-to-peer messages to create a distributed IoT network. ADEPT can find its software in a smart home, which can, for instance, become a semiautonomous unit for self-service and maintenance with a Samsung washing machine. If the machine is out of control, the operator is told about the malfunction and the installation of software upgrades on its own. The washing machine can connect with other intelligent devices in a network with ADEPT to maximize power efficiency. For example, if the TV is on, it may delay a washing cycle for several hours. It also helps the machine to monitor the stock of detergent used to pay for the order itself and obtain a confirmation of the retailer's delivery. The owner of the washing machine will then be told of the purchase on the smartphone. This is less futuristic than it might be, and ADEPT-style systems will soon win the market.

Chronicled, a self-reported IoT and blockchain laboratory, applied the pharmaceutical supply chains using a combination of these two technologies. The solution developed helps pharmaceutical producers, wholesalers, and hospitals to track every step of drug transportation and makes it impossible for criminals to unload stolen drugs.

7.13 CONCLUSION

IoT-based organizations face challenges in recognizing security technologies and methods necessary to mitigate specific IoT threats to IoTs. Technology from blockchain aims to play an essential role in meeting these challenges. Niche security vendors will begin to provide these services, but the integrity and authenticity services offered by blockchain applications can be taken advantage of immediately. We highlighted features in this chapter while trying to protect connected devices using blockchain technology. However, because of IoT hardware limitations, we infer that several of them cannot function as a transaction node and will, thus, be outside the stable blockchain within the scope of several hundred thousand or more IoT devices. Many devices can benefit from blockchain services' protection and other features through upstream network transaction nodes and specialized intermediaries' APIs. These upstream features can be used to secure IoT devices (configuring and upgrading power, secure firmware updates) and communications (IoT discovery, trustworthy contact, authentication/signing of the message). We hope that this chapter encourages business leaders and entrepreneurs, taking advantage of the chance to broaden this technology's ability to protect the IoT.

REFERENCES

Ai, Y., Peng, M., & Zhang, K. (2018). Edge computing technologies for Internet of Things: a primer. *Digital Communications and Networks*, *4*(2), 77–86. https://doi.org/10.1016/j.dcan.2017.07.001

Allhoff, F., & Henschke, A. (2018). The Internet of Things: Foundational ethical issues. *Internet of Things*, *1–2*, 55–66. https://doi.org/10.1016/j.iot.2018.08.005

Alotaibi, B. (2019). Utilizing blockchain to overcome cyber security concerns in the Internet of Things: A review. *IEEE Sensors Journal*, *19*(23), 10953–10971. https://doi.org/10.1109/JSEN.2019.2935035

Atlam, H. F., Azad, M. A., Alzahrani, A. G., & Wills, G. (2020). A review of blockchain in internet of things and AI. *Big Data and Cognitive Computing*, *4*(4), 1–27. https://doi.org/10.3390/bdcc4040028

Atlam, H. F., & Wills, G. B. (2020). IoT security, privacy, safety and ethics. In *Internet of Things* (Issue August). Springer International Publishing. https://doi.org/10.1007/978-3-030-18732-3_8

Banerjee, M., Lee, J., & Choo, K. K. R. (2018). A blockchain future for internet of things security: A position paper. *Digital Communications and Networks*, *4*(3), 149–160. https://doi.org/10.1016/j.dcan.2017.10.006

Brous, P., Janssen, M., & Herder, P. (2020). The dual effects of the Internet of Things (IoT): A systematic review of the benefits and risks of IoT adoption by organizations. *International Journal of Information Management*, *51*(May 2019), 101952. https://doi.org/10.1016/j.ijinfomgt.2019.05.008

Butun, I., Osterberg, P., & Song, H. (2020). Security of the Internet of Things: Vulnerabilities, attacks, and countermeasures. *IEEE Communications Surveys and Tutorials*, *22*(1), 616–644. https://doi.org/10.1109/COMST.2019.2953364

Compton, J. (2017). How blockchain could revolutionize the Internet of Things. *Forbes*, 1–5. www.forbes.com/sites/delltechnologies/2017/06/27/how-blockchain-could-revolutionize-the-internet-of-things/#8defb766eab4

Fawaz, K., & Shin, K. G. (2019). Security and privacy in the Internet of Things. *Computer,52*(4), 40–49. https://doi.org/10.1109/MC.2018.2888765

Ghosh, A., Chakraborty, D., & Law, A. (2018). Artificial intelligence in Internet of Things. *IET*, 1–11.

Iansiti, M., & Lakhani, K. R. (2017, January–February). The truth about blockchain. *Harvard Business Review,2017.*

Jurcut, A., Niculcea, T., Ranaweera, P., & Le Khac, N. A. (2020). Security considerations for Internet of Things: A survey.*ArXiv*, 1–19. https://doi.org/10.1007/s42979-020-00201-3

Khanna, A., & Kaur, S. (2020). Internet of Things (IoT), applications and challenges: A comprehensive review. *Wireless Personal Communications,114*(2). Springer US. https://doi.org/10.1007/s11277-020-07446-4

Khvoynitskaya, S. (2020). *Blockchain for IoT security—a perfect match Convenience is prioritized o er sec rit.* Itransition. www.itransition.com/blog/blockchain-iot-security

Liau, C. H., Shen, W. W., & Su, K. P. (2006). Towards a definition of the Internet of Things (IoT).*IEEE Internet of Things*, *60*(1), 121–122.

discovery. *Proceedings of the International Conference on Parallel and Distributed Systems—ICPADS, 2017-Decem* (January), 360–367. https://doi.org/10.1109/ICPADS.2017.00055

Namasudra, S., Deka, G. C., Johri, P., Hosseinpour, M., & Gandomi, A. H. (2020). The revolution of blockchain: State-of-the-art and research challenges. *Archives of Computational Methods in Engineering, 0123456789.* https://doi.org/10.1007/s11831-020-09426-0

Palo Alto. (2020). 2020 unit 42 IoT threat report. *Palo Alto.* https://drive.google.com/open?id=1VLA1IweXyJMVeWxvy_8vwtypUQXB_Uhn

Patel, K. K., Patel, S. M., & Scholar, P. G. (2016). Internet of Things-IoT: Definition, characteristics, architecture, enabling technologies, application & future challenges. *International Journal of Engineering Science and Computing*, *6*(5), 1–10. https://doi.org/10.4010/2016.1482

Rainie, L., & Anderson, J. (2017, June 6). The Internet of Things connectivity binge: What are the implications? *Pew Research Center*, 1–94. www.pewinternet.org/2017/06/06/the-internet-of-things-connectivity-binge-what-are-the-implications/

Rejeb, A., Keogh, J. G., & Treiblmaier, H. (2019). Leveraging the Internet of Things and blockchain technology in Supply Chain Management. *Future Internet*, *11*(7), 1–22. https://doi.org/10.3390/fi11070161

Russell, B., Lingenfelter, D., Abhiraj, K. S., Manfredi, A., Anderson, G., Mordeno, A., Bell, M., Mukherjee, V., Bhat, G., Naslund, M., Boyce, K., Nieto, J., Cook, M., Owen, T., De Monts, R., Rastogi, A., Donahoe, T., Sanchidrian, G., Drake, C., . . . Thriveni, S. (2015, April). Security guidance for early adopters of the Internet of Things (IoT). *Mobile Working Group Peer Reviewed Document*, 1–54. https://cloudsecurityalliance.org/research/surveys/

Salim, M. M., Rathore, S., & Park, J. H. (2020). Distributed denial of service attacks and its defenses in IoT: A survey. *Journal of Supercomputing,76*(7). Springer US. https://doi.org/10.1007/s11227-019-02945-z

Sreevani, S. (2019). *Blockchain to secure IoT data.* www.geeksforgeeks.org/blockchain-to-secure-iot-data/

Tawalbeh, L., Muheidat, F., Tawalbeh, M., & Quwaider, M. (2020). IoT privacy and security : challenges and solutions. *Applied Sciences—Open Access Journal, 10*(4102), 1–17.

Trend Micro. (2018). *Blockchain: The missing link between security and the IoT?* www.trendmicro.com/vinfo/mx/security/news/internet-of-things/blockchain-the-missing-link-between-security-and-the-iot

Ud Din, I., Guizani, M., Hassan, S., Kim, B. S., Khurram Khan, M., Atiquzzaman, M., & Ahmed, S. H. (2019). The Internet of Things: A review of enabled technologies and future challenges. *IEEE Access*, *7*, 7606–7640. https://doi.org/10.1109/ACCESS.2018.2886601

Zhang, R., Xue, R., & Liu, L. (2019). Security and privacy on blockchain. *ArXiv*, *1*(1).

8 Spear Phishing Detection

An Ensemble LearningApproach

Shibayan Mondal, Samrajnee Ghosh, Achiket Kumar, SK Hafizul Islam, and Rajdeep Chatterjee

CONTENTS

8.1 INTRODUCTION

Phishing has become one of the biggest threats to Internet users. Phishing is an Internet attack in which an adversary spots victims and steals valuable information from them. This sensitive data can disclose victims' financial or personal information and then fraud is commited. Sometimes, attackers also use this information to hack into the victims' systems and further spread the attack. Most phishing attacks are

DOI: 10.1201/9781003152392-8

made with the help of emails and websites and, of late, through social media posts. These emails or posts mostly look legitimate and, in fact, wildly urging so much that it is hard to know if they are harmful. Still, inside, they contain URLs, which can lead the victim to malicious websites that are rat traps for stealing personal information (Natekin et al., 2013). Sometimes, these emails or posts also contain malware whose installation can lead to severe information leakage. So, no matter how many firewalls, certificates, encryption software, or two-factor authentication mechanisms an organization applies, the person behind the keyboard can be a victim of a phish (Vishwanath et al., 2011). Some historical phishing attacks may give a better insight into phishing and the large-scale damage that it can bring.

8.2 HISTORY OF PHISHING

Phishing scams use spoofed emails, posts, and websites to tempt people in falling into their traps, just like fishing, where the fish are caught using baits. The name phishing was coined from the word "fishing" where the letter "f" was replaced by "ph" to refer to "phreaks," one of the earliest groups of hackers on the Internet. The term phishing was used first on January 2, 1996, in the Usenet newsgroup, AOHell, as per record. At that time, America Online (AOL), being one of the best Internet access providers, often attracted a host of hackers and pirated software sellers. These people eventually formed the "warez" community, believed to have been the first phishing attacker community. It used to steal people's passwords and used algorithms to generate random credit card numbers. Later, these credit card numbers were used for logging in to people's AOL accounts and stealing information from there. The company finally banned this practice in 1995 when AOL developed some security measures so that they could prevent the generation of random credit card numbers and their successful application. With this racket being shut, the phishers resorted to instant messaging AOL users by disguising themselves as AOL employees. They used to ask the users to verify their accounts or billing information confirmation provided by them. Such attacks were very new and innovative, and people often fell for their traps. The phishers even opened up AIM accounts which were not punishable by AOL TOS department. Later, AOL was compelled to include warning messages in their emails to alert their users.

Attackers have not changed their methods a lot. In 2001, the phishers' primary attention was on online payment systems. Their first such attack was on E-Gold in June 2001 although it was not considered very successful. In 2003, the attackers started registering many domains for different websites, which were built to match famous e-commerce websites like eBay. If the users did not pay enough attention, they would often fall into the hole dug by these phishers. Some phishers also used email worms to send spoofed emails that contained URLs leading to fake PayPal sites where customers were made to update their credit card details and other identifying information. By 2004, the phishers had become very successful as they were successfully attacking banking sites using pop-up windows to steal sensitive information from the customers.

In late 2008, Bitcoin as well as other cryptocurrencieswere launched where transactions using malicious software were made anonymous, and this changed the game for cybercriminals. Phishing eventually became an official part of the black market. The first cryptographic malware spread was staged by Cryptolocker in September of 2013 when malware was spread by downloads from a compromised website. It was then sent to victims in the form of two different kinds of phishing emails. The emails had zip files disguised as customer complaint records and malicious links with messages regarding a problem clearing a check. Once clicked, Cryptolocker scrambled and locked the computer files and demanded the owner to make a payment for the key to get the files unlocked and decrypted.

Lately, phishers have started adopting HTTPS more and more often on their sites, starting in 2017. Recently, the news of the Twitter attack of July 15, 2020 had taken over the Internet by storm, and many oversized shirts were victims of this infamous attack. What is more, it is a social platform as trusted as Twitter falling prey to this awful attack described as a "phone spear phishing attack." Once the hackers gained access to an internal dashboard meant for Twitter employees, they took control of the accounts. They appeared to have used social engineering to gain access to the tools. The tool, which Motherboard first reported, apparently allowed hackers to take the authority of the accounts by changing associated email addresses without informing their owners. It was a social engineering attack (phishing); 130 accounts were hacked, including that of Bill Gates, Jeff Bezos, Elon Musk, and so on, and used to solicit cryptocurrency (Sarkar et al., 2019).

8.3 STATISTICS OF PHISHING

According to the FBI, one of the most common types of cybercrime in 2020 was phishing. Phishing cases became more and more familiar with a boom in the number of instances from 114,702 cases in 2019 to 241,324 cases in 2020 (Yeboah-Boateng et al., 2014). Google safe browsing is trusted with detecting phishing sites on the Internet, and it has registered 2,145,013 phishing sites as of January 17, 2021. This is a significant hike compared to the malware sites, which are about 28,803 in number at the same time. Figure 8.1 shows the comparison of an increasing rate of unsafe websites between 2007 and 2019. According to the recent ESET threat report, in Q3 of 2020, phishing emails containing the most common types of malicious files were as follows:

1. Windows executables files—74%
2. Script files—11%
3. Office documents—5%
4. Compressed archives—4%
5. PDF documents—2%
6. Java files—2%
7. Batch files—2%
8. Shortcuts—2%
9. Android executables—>1%

Even the statistics of phishing in recent times have been quite alarming. It is evident that with digitization in every sector, the threat of phishing and the variety of methods for staging it are also increasing. Some of the statistics about phishing are presented in the following:

1. Phishing attacks reportedly account for more than 80% of the security incidents.
2. Due to phishing, 17,700 dollars is lost every minute.

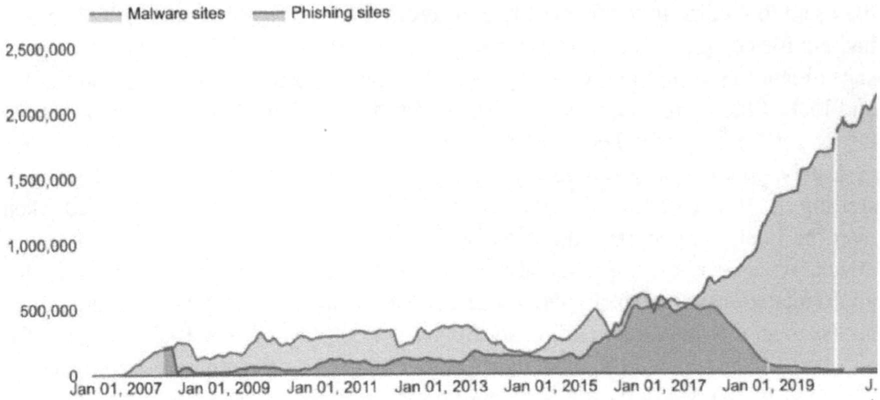

FIGURE 8.1 Rise in the number of unsafe websites between January 2016 and January 2021.

Source: Pulled from Google Safe Browsing (2021)

3. Sixty percent of the breaches involved vulnerability for which a patch was available but not applied.
4. Ninety-seven percent of the people cannot identify a phishing scam.
5. As high as 77%, IT professionals acknowledge that their teams are not prepared to face cybersecurity challenge today.
6. Nearly 1.5 million new phishing sites are created every month (Ferreira et al., 2015).

As is customary in Q1, China (15.82%) and the United States (12.64%) were among the top spam-originating countries. Coming to the other top three, Germany was in the fifth place in Q1 2019 (5.86%), ceding Russia (6.98%) to the third place, and allowing Brazil (6.95%) to steal the fourth. Then, France (4.26%) came to the sixth place. Argentina (3.42%), Poland (3.36%), and India (2.58%) are in the seventh, eighth, and ninth positions, respectively. The tenth position was acquired by Vietnam (2.18%) (Bonaccorso et al., 2017).Figure 8.2 shows the country-wise comparison in terms of originating spam.

8.4 ANATOMY OF AN ATTACK

Attackers usually follow a primary procedure when they are entrapping victims. The anatomy of an attack can be divided into five steps as follows:

1. *Collect*: In this step, the phishers gather information about a specific target by exploiting all available resources on social media to make the attack more successful. The phishers collect personal information, *i.e.*, email addresses, phone numbers, interests, and educational or working history. For example, in the 2017 French election, phishers collected information about Macron's campaign. Facebook took action and blocked 24 such profiles collecting information. On a later investigation, they were found to be the same attackers' group, "Fancy Bear," who also tried to hack the American elections.

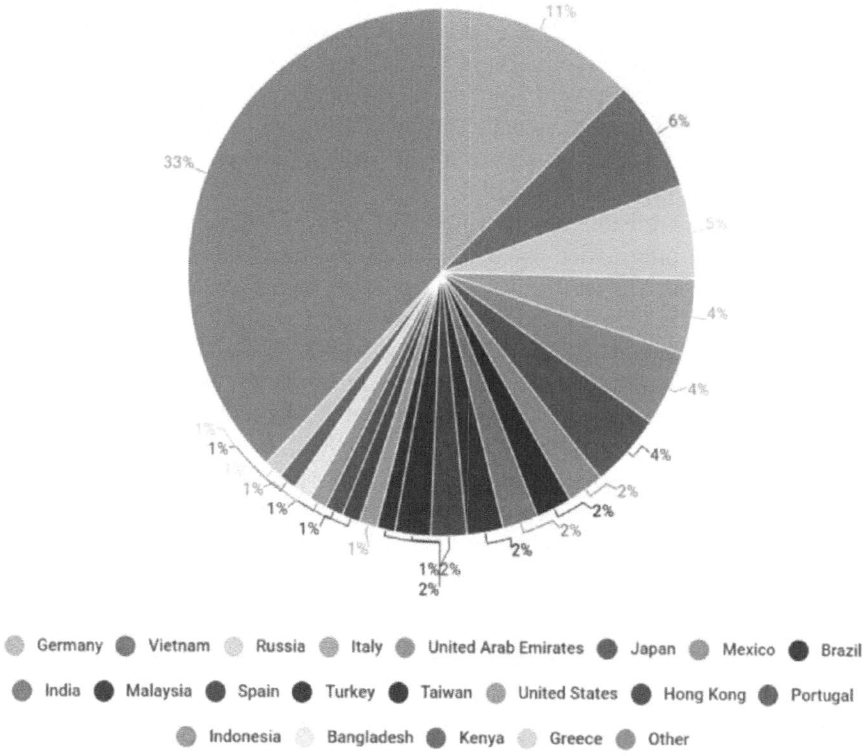

FIGURE 8.2 Graphical representation of the percentage of spam originated from various countries (top 20).

Source: Kaspersky Lab (2019)

2. *Construct:* In this step, the attackers try to get involved with their victims by creating fake social media accounts and communicating with them regularly. This is the most crucial aspect of social engineering attacks being successful—social approaches. Now attackers frequently use sociopsychological techniques,*i.e.*, Cialdini's principles of persuasion to manipulate their victims (Hosmer et al., 2000). The new character that the attacker pretends to have fabricates credentials and usually establishes a common ground with the victim and many times with the same work organization or university. In February 2017, SecureWorks helped a middle-eastern company to diagnose a security event where a spyware infection was attempted. The attack was launched by Iranian hackers who created fake LinkedIn, Facebook, Twitter, and Instagram accounts.
3. *Contact:* In this step, targets are contacted. When the victim falls for it, the attacker can easily access nonpublic information about the victim. It is usually done through emails or by sending friend requests on social media. It is through these ways that infected hyperlinks or other malware are sent

to the target. In the American elections of 2016, more than 30 Clinton staffers received spear-phishing emails from hackers who created fake email accounts for this purpose. These emails contained links to a website that hackers operated. In April 2016, the accessed information was used against the Democratic Congressional Campaign Committee to hack into their system so that data could be stolen with the use of malware. All in all, 33 DNC systems got hacked in this way.

4. *Compromise*: It is the step in which malware is installed in a victim's system so that information could be stolen from the system, and it could be compromised. The Isis-affiliated group Cyber Caliphate did this very same thing in 2015. They hacked the US central command employee's YouTube and Twitter accounts to spread their propaganda (Christine et al., 2004).
5. *Contagion:* It is the last step of the model. In this step, attackers take advantage of the hacked account to enlarge the attack by targeting new victims. This way, the attack spreads from one system to many systems. This step is usually used to magnify the outcome and mass spread the attack. It is pretty well said that "Contagion is especially dangerous because threat actors can target vulnerable victims and scale up to bigger targets."

8.5 KINDS OF PHISHING

Phishing is just a broad group of frauds usually spread through emails or social media but based on the platform used for the attack. Phishing can be classified into the following kinds:

1. *Emailing*: Email phishing has been prevalent even in the 1990s. Hackers send malicious emails to a list of email IDs they collect. Usually, the phishers' first step is to set up a fraud website. After that, they send a large volume of fake emails containing links to that maleficent website. When victims click on the attached URL, they are headed to a website that demands their personal information. Most of the emails are very compelling and usually talk about a compromise with the victim's account. They also console the victim by saying that it can be undone if the victim goes to the attached link and fills out the necessary information. It often scares the victims, and they immediately do as suggested. Thus, the phishers gain access to victims' sensitive details and later use them to achieve financial benefits from different online transaction sites. Usually, phishing emails are easy to spot because English used in those emails is not grammatically sound. These emails typically sound choppy because they go through translation programs in many different languages before arriving in English. Overtime, phishers are honing their skills of committing frauds, so they now carefully craft their emails in English and as a result they become indistinguishable. Another emailing scam is called sextortion, where the attackers write emails that claim that they have access to the victim's system and know the victim's passwords. The main feature of these scams is that the attackers say they have tracked the victim's Internet activities,

including some unsocial elements. The attackers threaten to leak them to the victim's friends and family if the victim does not provide them with the demanded ransom.

2. *Social media phishing*: With the increasing popularity of social media platforms like Instagram, Facebook, Twitter, and LinkedIn, phishing on these platforms is also becoming inevitable. In modern times, businesses have also hopped onto social media sites to increase their reach. It is used as an opportunity by bad actors for whom phishing on such sites has become a matter of a few clicks with tools like "Hidden Eye" or "ShellPhish." The hackers often aim to steal login details of social media accounts, bank details, and private information from the targets. Social media phishing can again be divided into Instagram Phishing, LinkedIn Phishing, Twitter Phishing, Facebook Phishing, and so on. These are described as follows:

    a. *Instagram phishing*: Instagram is an online platform where texts and photos are shared, and it has a large population as of now. A phishing attack for this platform begins with the attacker forming a fake login website that has been carefully crafted to match the login page of Instagram itself. The users get fooled by the appearance of these sites and enter their login credentials which the bad actor then captures. The user is then redirected to the actual Instagram login page. The attacker with the users' credentials now has full access to their accounts. It is very dangerous because some users have the same login information for all of their social media accounts and, worse still, their bank accounts for ease of access. The hacker can also commit identity theft by changing the hacked user's details and preferences or asking for personal information from friends and family or, worse, by changing the password and logging out the real user from accessing the account.

    b. *LinkedIn phishing*: LinkedIn is a platform where professionals can connect with each other and form a network. Hackers attack by sending emails, LinkedIn messages, and links to the users who attempt to steal information like login details, bank details, and other private information. After a successful attack, the hacker can use the prey's LinkedIn account to convince people in his community by sharing helpful posts and sales to retrieve personal information. The attacker can also design messages in such a way that they seem to be coming from the LinkedIn authority. Attacks on LinkedIn are easier to stage because the official LinkedIn site uses many email domains. So, it becomes easy for the attacker to pose as an absolute LinkedIn authority because keeping track of all the official domains in use is very difficult.

    c. *Twitter phishing*: Twitter is a platform where small messages, news, or photographs could be exchanged between people or shared in a community. With its character restriction to 280, users on this platform can quickly scan through the messages. Hacking on Twitter occurs similar to other social media sites—the hackers send emails and messages and lure the users into falling into their traps, thus stealing valuable information from them. After these attacks, other related attacks occur, one of them being "Pay for followers." In this kind of attack, the hackers claim to guarantee a large pool of followers if the user pays a minimal sum of about five dollars. If the user takes the bait, he would probably fall prey to the attacker's evil intentions, and his account might also become a tool to carry out further attacks.

   d. *Facebook phishing*: Facebook was launched in the early 2000s, and many people use it for communicating with their friends and family. Facebook phishing usually begins with a fake message from the hacker saying that something has gone wrong with the user's Facebook account and that they should mend it by going to the URL attached with the message. The URL is a path to an easy-to-follow, Facebook lookalike fraud website designed by the impostor. If the user follows the link and provides the details there, the attacker would capture all the information and use it to spread the spam.

3. *Smishing:* Smishing (James et al., 2006) uses the text message service, which is also known as SMS. It is because many people are more likely to trust text messages as compared to emails. People think it is difficult for hackers to get their phone numbers as compared to email IDs. Still, truth being said—it is easier to get hold of the phone number because there are only a finite number of options with phone numbers given in a particular country as they have a fixed number of digits. In contrast, email IDs can vary in length, including a wide variety of characters. According to Gartner's reports, 98% of text messages are read, and 45% are responded to, whereas only 6% of emails are responded to, and, so, texts are a very safe place for hackers to lay their trap in. The attackers sometimes pretend to be bank employees and ask for personal details from the victims. In the texts, they sometimes ask victims to click on links that are supposed to connect them to the bank's website which instead connect them to some malicious lookalike website developed by the hacker. Sometimes, victims are also asked to call the bank's customer service number conveniently provided with the text to clear out some recent suspicious charges on their account. Another method is to play with human sympathy and ask for charity for the needy from the victims. In such cases, the victim has to go through a link provided with the text message and provide all the credit card details, which the attacker then captures. Another method of smishing that is very common nowadays is a message from the phone carrier offering the victims an incredible deal on a service or phone upgrade. Following the link to receive the offer can land the victim on a fraud website which tricks the victim into revealing all personal information.
4. *Search engine phishing:* This is a very common and vicious kind of phishing. It is also known as SEO Trojan. In such cases, the hacker tries to be one of the top results when using the search service by Google, Bing, or other search engines. If the attacker gets victims to click on his website link, they are trapped, and all their data is stolen from them.
5. *Vishing*: Vishing has very similar motives as the other phishing attacks, but it occurs on voice calls. Usually, in vishing attacks, the attacker pretends to be some software company employee on the call and reports the presence of some virus in the victim's system. The attacker then suggests an antivirus and asks

for credit card information for buying the same. Thus, the victim gets trapped, and the attacker withdraws large sums of money from the victim's account.

Phishing can be committed with many motives in mind. Some of them include attacking a particular group of people or an individual of high social standing to make substantial financial benefits from them. Based on the attacks' scale and nature, phishing can be divided into the following categories.

1. *Spear phishing*: Spear phishing mainly targets a group or type of people such as some organization's important employees or representatives. This kind of phishing is very well choreographed because the attackers target specific people and hence do good research on their backgrounds which aids them in framing more meaningful conversations that can easily trap the victims. Spear phishing was started by sending spam emails, but gradually, with the advent of social media, spear phishing occurred through social media. It is because most of the information that the attackers want is openly available on their social media accounts, like the victims' preferences, connections, and backgrounds. Thus, spear phishing has now become very easy to stage.
2. *Whaling*: Whaling is a little more specific kind of phishing because it points its hook at the whales, i.e., the bigger fish in the pond. Whaling basically targets prominent faces from the industry, such as CEOs, CFOs, and other people of equivalent eminence. Whaling emails mostly scare the victims by saying the company is getting sued, and the victim needs to click on a particular link for more information. The link then directs to a website that asks for critical information such as tax ID and bank account details.

8.6 DATASET

In recent times, spear-phishing attacks on social media have seen quite a rise. In fact, according to a study in 2019, about 88% dealt with spear-phishing attacks, and 86% experienced social media phishing. So, the rest of this chapter mainly focuses on detecting or predicting spam in social media using different machine learning (ML) methods.

So, since the aim is to apply ML, there has to be some training done with previous such spam cases in social media to predict future spam. The dataset (Vergelis et al., 2019) that this chapter focuses on is a collection of 14,899 tweets, and it has the following features:

- *Tweet*: The column contains the whole text body of the tweet along with any tagging done to any individual.
- *Following*: It gives the total count of the number of accounts that are being followed.
- *Followers*: It gives the total count of the number of accounts that are following.
- *Actions*: It denotes the total number of favorites, replies, and retweets of the said tweet.

- *is_retweet*: In this column, the binary [0,1] value is given; if the value is 0, then it is not a retweet; otherwise, it is a retweet.
- *Location*: It gives the self-written location provided by the user on their profile, which may not exist, be "unknown," and is not standardized, for example, could be "NY," "New York," "Upper East Side," etc.
- *Type*: It states whether the said tweet is spam or not.
- *Type_num*: This column is the numerical version of the type column; it gives 1 for spam and 0 for quality.

As seen in Table 8.1, the first few data from the actual dataset are given. As already mentioned, spear phishing targets people with good fame and influence in society. The frauds are large scale, and the eminent trustworthy person can unknowingly transmit or spread the attack to many people who trust them. So, such big shots would have a large

TABLE 8.1
Description of the Features of the Dataset with Example

Tweet	Following	Followers	Actions	is_retweet	Location	Type
Good Morning Love @ LeeBrown_V	0	0	0	0	Pennsylvania, USA	Quality
'@realDonaldTrump @ USNavy RIP TO HEROES'	42,096	61,060	5,001	0	South Padre Island, Texas	Spam
Haven't been following the news, but I understand #EFF was doing the dumbest things	0	0	NaN	0	Will never be broke ever again	Quality
pic.twitter.com/dy9q4ftLhZ What to do with paper scissors and glue http://paperlandmarks.com/product/the-parthenon-paper-model-kit . . . #papercraft #diy	0	0	0	0	Mundo	Quality
#DidYouKnow ► Mahatma Gandhi made a brief visit to lecture in #Nottingham on October 17, 1931 [@MumblingNerd]	17,800	35,100	NaN	0	Nottingham, England	Quality

Source: Based on UtkMl's Twitter Spam Detection Competition (Kaggle 2019)

follower count on Twitter. So, to boil down the dataset to focus on the famous people, the dataset was filtered, and only the tweets tweeted by accounts having a follower count of greater than 25,000 were kept. It ensured that most of the tweets came from well-known accounts that had a good hold on society, and it also ensured that these accounts mainly were "verified" by Twitter. After the filtration and cleaning of unwanted data, the dataset looked like as given in Table 8.2. The follower's column has values greater than or equal to 25,000 only, and any other tweet has been pruned from the dataset leaving behind a filtered dataset with 1,985 tweets. Next, to correspond between the text of the tweets and a numerical form to represent them so that they become fit for use in the ML methods, the Count Vectorizer method is used. Count Vectorizer is a method for extracting features from a dataset, and it is used to convert a large corpus of text documents to an array or rather vector that gives the count of terms or tokens. Thus, it is a very flexible method to represent features. So, a two-dimensional array of all possible words used in the tweets is formed where for every tweet, the value of the array vector would be 0 if the corresponding word is absent from the tweet and 1 if it is present.

8.7 MACHINE LEARNING

ML can be deployed to spot phishing and learn its patterns. ML also has another advantage—several of its methods can be combined in different ways to form a

TABLE 8.2
Filtered Dataset Displaying the Tweets Having Followers More than 25,000

Tweet	Following	Followers	Actions	is_retweet	Location	Type	Type_num
@realDonaldTrump @USNavy RIP TO HEROES	42,096.0	61,060.0	5,001.0	0.0	South Padre Island, Texas	Spam	1
#DidYouKnow ► Mahatma Gandhi made a brief visit to lecture in #Nottingham on October 17, 1931 [@MumblingNerd]	17,800	35,100	NaN	0	Nottingham, England	Quality	0
Please don't talk about me like that I'm only a little tipsy	0.0	6,200,000.0	NaN	1.0	Los Angeles, CA	Quality	0
Need I remind everyone that the first lines you learn to say as a rapper coming up are "I am the best"	0.0	275,000.0	80.0	0.0	Johannesburg, South Africa	Quality	0

better model that more successfully classifies information—this method is called ensemble learning, and it is explained in detail later on (Cunningham et al., 2021).

8.7.1 Basic Machine Learning Algorithms

First, some basic and popular ML algorithms are used with the chosen dataset to see their accuracy in detecting spear phishing. Later, these methods are improved upon by combining them with the help of different ensemble learning methods.

8.7.1.1 Support Vector Machine

SVMs are statistical tools based on supervised learning, which are used for classifying data and are very robust (Abu-Nimeh et al., 2007). SVM can be classified into two different types—linear SVM and nonlinear SVM. Linear SVMs separate a collection of data points into classes by using a linear hyperplane, whereas nonlinear SVMs do the same, but with nonlinearly separable data points by transforming the feature space they are in.

Linear SVM: Let us consider n data points as $(x_1, y_1), (x_2, y_2), \ldots, (x_n, y_n)$, where each y_i is a 1 or a −1 telling where x_i belongs. Here, x_i is a real vector of p dimension. The idea is to compute the "maximum margin hyperplane" such that it separates the group of points where $y_i = 1$ from the group of points where $y_i = -1$. In this way, the distance between the plane and the nearest point x_i from any of the two groups gets maximized. A hyperplane can be expressed in the following way:

$$\mathbf{w}^{\mathrm{T}}\mathbf{x} - b = 0$$

Here, w is the normal vector to the hyperplane. If the dataset can be separated linearly, then two parallel hyperplanes can be chosen such that they separate the two classes of data, and the distance between the two classes is also maximized. The area between these two hyperplanes is called the *margin*. The maximum margin hyperplane is considered midway between these two hyperplanes. With a normalized dataset, these hyperplanes can be expressed as follows:

$\mathbf{w}^{\mathrm{T}}\mathbf{x} - b = 1$ (anything on or above this hyperplane comes in the class with label 1)
$\mathbf{w}^{\mathrm{T}}\mathbf{x} - b = -1$ (anything on or below this hyperplane comes in the class with label −1)

This is called hard margin. Also, to restrict data points from being inside the margin, the following constraints are used:

$$\mathbf{w}^{\mathrm{T}}\mathbf{x}_i - b \geq 1 \text{ if } y_i = 1 \text{ or } w^{\mathrm{T}}x_i - b \leq -1 \text{ if } y_i = -1$$

These two equations can be combined to be written as:

$$y_i(\mathbf{w}^{\mathrm{T}}\mathbf{x}_i - b) \geq 1 \text{ for all } 1 \leq i \leq n$$

- **Nonlinear SVM:** To extend the use of SVM to nonlinearly separable data, the "hinge loss" function is used, which is stated as follows:

$$\mathbf{max(0, 1—}y_i \, (w^T x_i\_b\,))$$

If x_i lies on the right side of the margin, then this function gives zero, whereas if x_i is on the wrong side of the margin, then the value given by the function is proportional to the distance from the margin. The nonlinear classification is achieved by applying the "kernel trick" on the linear classifier's maximum margin hyperplane. In this technique, every dot product is replaced by a nonlinear kernel function that allows the program to fit the greatest margin hyperplane in a modified feature space. It frequently leads to a nonlinear transformation and a changing space with a high number of dimensions. It indicates that the separator may be nonlinear in its original input space despite being linear in the converted space. Polynomial homogeneous, polynomial nonhomogeneous, Gaussian radial basis function, hyperbolic tangent, and other standard kernels are utilized in nonlinear SVMs.

Using SVMs without any kernel trick gave an accuracy of about 93.12%. The pros and cons of using SVM are listed in the following.

Advantages:

- SVMs work very well with a clear margin of separation.
- They are very effective in transformed high-dimensional feature spaces.
- They are also very efficient in cases where the degree of dimension is higher than the number of samples.
- They use only a few data points for training purposes. These points are called support vectors. So, they are memory-efficient too.

Disadvantages:

- They do not perform very well with larger datasets, because the training time required in such cases is comparatively higher.
- They also do not perform very well when there is a lot of noise in the data, i.e., when the target functions overlap.
- SVMs do not immediately offer the probability as an output; instead, it is computed via a time-consuming fivefold cross-validation approach.

8.7.1.2 Decision Tree

A decision tree is a decision-making aid that employs a tree-like model of decisions and their probable consequences, such as chance event outcomes, resource costs, and utility (Safavian et al., 1991). Each internal node represents a test of a single property in a flowchart-like layout. The test results are represented by the branches that emerge from those nodes, and each leaf node represents a class label. As a result, the categorization criteria are represented by the pathways from root to leaf. A decision tree now includes three different sorts of nodes.

- *Decision node*: A square represents a node in a decision. The decision maker selects an action in a decision node, i.e., one of the several edges (at least two edges come out of a decision node) stemming from that node. A strategy for the entire tree unanimously prescribes an action in every decision node.
- *Chance node*: A circle represents a chance node in a decision tree. In a chance node, one of the edges stemming from it representing a reaction is randomly selected. Each edge emerging from a chance node is associated with a particular probability. The probability represents the chances of occurrence of the reaction associated with the particular edge of the chance node.
- *Terminal node*: A triangle represents a terminal node in a decision tree. Terminal nodes determine the end of a sequence of actions and reactions in a decision problem.

In a decision tree, the entropy of the model is calculated by the following formula.

$$E(S)=\sum_{i=1}^{n} - p_i \log_n p_i$$

where S is the sample, p_i is the probability of occurrence of a class label i as the output, and n is the number of the class labels that are possible as the output. Some properties of this calculated entropy are stated as follows:

- The entropy becomes 1, i.e., $E(S) = 1$ if the classes are equally distributed.
- The entropy becomes 0, i.e., $E(S) = 0$ if one class completely dominates over the other, i.e., for a completely homogeneous sample, i.e., $E(S) = 0$.
- For Gaussian distribution of data, the entropy is high, but it is still low as compared to uniform distribution.
- In case of uniform distribution, the entropy, i.e., $E(S)$ is maximum because, in uniform distribution, all have equal probability.
- For peaked distribution, the value of entropy is low, close to 0, because the data is unequally distributed.

There is another measure for impurity known as the Gini impurity. The formula for measuring the same is given as follows:

$$Gini=\sum_{i=1}^{n} p_i^2$$

where p_i is the probability of the class label *i*. *Gini impurity* is basically a close approximation of the entropy, and it is used majorly because it is computationally faster as calculation of square is faster than calculating the value for entropy which requires logarithmic as well as multiplicative calculations. Now, the calculation of entropy or *Gini impurity* is required to calculate the "*information gain*" of the nodes. It is given by the following formula:

$$IG(D_p, f) = I(D_p) - \frac{N_{left}}{N} I(D_{left}) - \frac{N_{right}}{N} I(D_{right})$$

where *IG* signifies the information gain, D_p denotes the dataset of the parent node, *f* is the feature considered for splitting, *I* indicates the impurity criterion, i.e., entropy or *Gini index* of the node, N_{left} signifies the number of samples for the left child, D_{left} indicates the dataset for the left child node, N_{right} denotes the number of samples for the right child, *N* denotes the number of samples for the parent node, and D_{right}signifies the dataset for the right child node. Now, in a decision tree whenever a node has to be chosen to form an action/reaction sequence, the value of *IG*is considered. The node with maximum *IG* is considered. If more than one node has the maximum *IG*, then all of them are considered.

The accuracy achieved with decision trees having a maximal depth of 1 in this dataset is 92.95%. Some pros and cons of using decision trees are as follows:

Advantages:

- These are very easy to understand and implement.
- They produce outcomes despite the lack of hard evidence. They also calculate the worst, best, and anticipated values in various circumstances, which aids in the acquisition of crucial information.
- They work flawlessly when combined with other decision techniques.

Disadvantages:

- They are very unstable, in the sense that even little changes in data can result in large changes in the structure of the optimum decision tree.
- If the data contains categorical variables with varying numbers of levels, the information gained by the decision tree is skewed toward the qualities with more levels.
- When several variables are unknown or numerous outcomes are connected, calculations can get exceedingly complicated.

8.7.1.3 Logistic Regression

Logistic regression is a supervised learning technique for solving classification issues. It's used to forecast the likelihood of a certain variable (Hosmer et al., 2000). Usually, the target variables used are dichotomous, which means only two possible classes are associated with them. So, the two classes are generally labeled using 1 (indicating success) and 0 (meaning failure). Mathematically, logistic regression predicts $P(Y = 1)$ as a function of *X*, meaning success. There are different types of logistic regression which are described as follows:

- *Binary/Binomial*: Usually, by logistic regression, it is referred to as binary regression. In these cases, as stated earlier, the target variables have two possible values, traditionally denoted by 1 or 0. These target variables can represent success or failure or, like in the case of the dataset discussed here—spam or ham.

- *Multinomial*: In this kind of classification, the dependent or target variable has more than two types of class labels. These labels are of the unordered type. The variables can categorize the data into categories like "A," "B," and "C.
- *Ordinal*: Here, the dependent or target variable has more than two types of ordered class labels. For example, these variables can represent grading systems like "Excellent," "Good," "Average," and "Poor," and each category can have scores like 3, 2, 1, and 0.

To explain logistic regression mathematically, a model with two predictors *x1* and *x2* and a binary response variable *Y* denoted by $p = P(Y = 1)$, is taken into consideration (Seymour et al., 2016). The assumption is that the predictor variables and the log-odds or logit of the event $Y = 1$ have a linear relationship. The following is a mathematical expression for the linear relationship:

$$l = log_b\left(\frac{p}{1-p}\right) = \beta_0 + \beta_1 x_1 + \beta_2 x_2$$

where l is the log-odd, b is the base of the logarithm, and β_i $(i = 0, 1, 2)$ are the parameters of the model. The odds can be recovered by exponentiating the log-odds as follows:

$$\frac{p}{1-p} = b^{(\beta_0 + \beta_1 x_1 + \beta_2 x_2)}$$

This equation by some simple algebraic manipulation becomes the following:

$$p = \frac{b^{(\beta_0 + \beta_1 x_1 + \beta_2 x_2)}}{b^{(\beta_0 + \beta_1 x_1 + \beta_2 x_2)} + 1}$$

On dividing this equation by $b^{(\beta_0 + \beta_1 x_1 + \beta_2 x_2)}$, the following result is achieved:

$$p = \frac{1}{1 + b^{-(\beta_0 + \beta_1 x_1 + \beta_2 x_2)}}$$

It basically translates to $S_b\left(\beta_0 + \beta_1 x_1 + \beta_2 x_2\right)$, which is a sigmoid function where b is the base. Thus, according to the previous formula, once the β_i are fixed, the probability that $Y = 1$ or the log-odds that $Y = 1$ for a given observation can be easily computed.

The implementation of logistic regression can be easily done using the Python sklearn.linear_model.LogisticRegression method. The parameters of the method are described as follows:

- *Penalty*: It is used to indicate the penalization standard. Only 12 penalties are supported by the solvers "newton-cg," "sag," and "lbfgs." Only the "saga" solver supports "elasticnet." No regularization is done if the value is "none" (the liblinear solver does not support it).
- *Dual*: The parameter is used to specify dual or primal formalization. Dual formulation is only implemented for the 12 penalties with a liblinear solver.
- *Tol*: It is the tolerance for stopping criteria.
- *Max_iter*: It specifies the maximum number of iterations needed for the solvers to converge.

The implementation of this method yielded an accuracy of 93.96% on the chosen dataset. The pros and cons of using this method are given hereafter:

Advantages:

- Logistic regression is one of the easiest and simplest supervised ML algorithms to implement. This is because it requires less computational power to be implemented than its counterparts and is very well suited for classification problems.
- The use of stochastic gradient descent in logistic regression makes it easier to update models to get or reflect new data, unlike the other methods.
- The probabilities resulting from this approach are well calibrated. This makes it more reliable than other models or methods that only give the final classification as results.

Disadvantages:

- In the case of high-dimensional models, overfitting becomes a problem. To curb this, regularization methods are used, but an over use of regularization techniques can result in underfitting and hence inaccurate results.
- Nonlinear problems cannot be solved using this method, and so the transformation of nonlinear problems into linear problems is needed, which can be time-consuming.
- This technique usually requires a higher number of observations; otherwise, it may result in overfitting the data.

8.7.1.4 Multinomial Naive Bayes

Naive Bayes classifiers are a type of probabilistic classifier based on the Bayes theorem and strong (Naive) independence assumptions between features in statistics. The number of parameters required for Naive Bayes classifiers is linear in the number of variables in a learning task, making them extremely scalable. Instead of the computationally costly iterative approximation approaches employed by other classification techniques, maximum-likelihood training may be done simply by evaluating a closed-form expression in linear time. There are three different kinds of Naive Bayes.

- *Gaussian Naive Bayes*: This method is particularly used in classification problems. It assumes that the features follow a normal distribution.

- *Multinomial Naive Bayes*: It is mostly used for discrete counts. For example, in a text classification problem, knowing the frequency of usage of words is very important for predicting or labeling the data, and this method provides just that.
- *Bernoulli Naive Bayes*: If the feature vectors are binary, i.e., in 0s and 1s, binomial models are useful. Text categorization with the "bag of words" paradigm, where the 1s and 0s represent "word exists in the document" and "word does not occur in the document," respectively, is an example of its use.

When there is discrete data, multinomial Naive Bayes is employed. The count of each word is used to predict the class or label in text learning. The feature vectors in a multinomial event model indicate the frequency with which particular events have been created by a multinomial ($p_1, p_2, \ldots, p_n$), where p_i is the chance that the ith event would occur. As a result, a feature vector $x = (x_1, x_2, \ldots, x_n)$ is essentially a histogram, with each x_i counting the number of times an event occurred in a certain instance. The odds of seeing a histogram are calculated as follows:

$$p(x \mid C_k) = \frac{\left(\sum_{i=1}^{n} x_i\right)!}{\prod_{i=1}^{n} x_i!} \prod_{i=1}^{n} p_{k_i}^{x_i}$$

where C_k represents the k possible classes and $p(x|C_k)$ represents the probability of x given C_k. When represented in log-space, the multinomial Naive Bayes classifier turns into a linear classifier as follows:

$$log\, p(C_k \mid x) \infty\, log\left(p(C_k) \prod_{i=1}^{n} p_{k_i}^{x_i} \right)$$

$$= log\left(p(C_k)\right) + \sum_{i=1}^{n} x_i log\left(p_{k_i}\right)$$

$$= b + w_k^T x$$

- where $b = log(p(C_k))$ and $wk_i = log(pk_i)$. For a given class and feature value in the training data, the frequency-based probability estimate will be zero. The probability estimate is directly proportional to the number of times a feature's value has occurred. This phenomenon creates a problem as all the information in other probabilities will be wiped out when multiplied. As a result, it is usual to include a small-sample correction, known as pseudocount, in all probability estimates to ensure that no probability is ever set to zero. When the pseudocount is one, the method of regularizing Naive Bayes is called Laplace smoothing, and in the general situation, it is called Lidstone smoothing.

The multinomial Naive Bayes method when applied on the chosen dataset gave an accuracy of 92.79%. The pros and cons of this method are as follows:

Advantages:

- A Naive Bayes classifier outperforms most other models when the assumption of independent classifiers is true.
- To estimate the test data, Naive Bayes requires a small amount of training data. As a result, the training period is shorter.
- Multinomial Naive Bayes is very easy to implement because only the probability has to be calculated.

Disadvantages:

- The assumption of independent predictors is the main flaw in Naive Bayes classifier. All of the attributes in Naive Bayes are assumed to be mutually independent. In reality, obtaining a set of predictors that are completely independent is nearly impossible.
- Assume that a categorical variable in the test dataset has a category that was not present in the training dataset. In that case, the model will assign a probability of 0 (zero) and will be unable to make a prediction. Zero frequency is another name for it.

8.7.1.5 K-Nearest Neighbor

The KNN method presupposes that the new case/data and existing cases are comparable. It assigns the new case to the category that is the most similar to the other options. The KNN algorithm stores all available information and classifies a new instance based on its similarity to the existing data. When fresh data emerges, the KNN algorithm can quickly classify it into a suitable category. So, in the case of classification problems, the output of this algorithm is a class membership (Mitchell, 1997). An item is classified by a majority vote of its neighbors. The item is placed in the class with the most members among its k closest neighbors (k is a positive integer, typically small). If $k = 1$, the item is allocated to that single nearest neighbor's class. The function is only approximated locally in KNN classification, and all computation is postponed until the function is evaluated. Because this method depends on distance for classification, normalizing the training data can substantially increase its performance if the features represent various physical units or arrive in radically different sizes. One of the first steps in implementing this algorithm is to choose the value of k. It is a very crucial step, and in order to do it, the following things should be kept in mind:

- *Using error curves*: Overfitting of data occurs with a very high variance if the value of k is small resulting in high test error and low training error. The error always happens to be zero if k =1 as the point itself is the nearest neighbor to that point. This is why high test error and low training error can be seen for low k values. It is called overfitting. As the value for k is increased, the test error is reduced. However, going beyond some specific k value introduces bias or underfitting, and hence test error increases. So, it can be said that the initial error in test data is high due to variance, gradually it becomes low, and then stabilizes. With a further increase in *k* value, it again starts to go up due to bias.

- Domain knowledge comes into play when choosing *k* value.
- When binary classification is considered, *k* value should be odd.

The pseudocode for this algorithm is stated as follows:

1. Load up the training data.
2. Prepare the data. Various methods such as scaling, dimensionality reduction, and missing value treatment are required to prepare the data.
3. Find the optimal value for *k*.
4. Predict a class value to generate new data
5. Calculate *distance*(X, X_i) for $i = 1, 2, 3, \ldots, n$, where X signifies the new data point, X_i denotes the training data, and *distance* indicates the chosen distance metric
6. Sort the distances with relevant training data in ascending order
7. Select the top "k" rows from this sorted list
8. From the given "k" rows, find the most common class. It provides us with the expected class.

The implementation of this method on the chosen dataset gives an accuracy of 87.92%. The advantages and disadvantages of this method are as follows:

Advantages:

- Lazy learner is the name given to KNN (instance-based learning). During the training phase, it does not learn anything. The training data isn't used to derive any discriminative functions. In other words, it does not require any training. It saves the training dataset and uses it only when making real-time predictions to learn from it. It makes the KNN method much faster than other training-based algorithms like SVM and linear regression.
- Because the KNN method does not require any training before making predictions, new data can be supplied without affecting the system's accuracy.
- KNN is a simple algorithm to use. The distance function and the value of *k* are the only two factors necessary to implement KNN (e.g., Euclidean or Manhattan)

Disadvantages:

- The cost of computing the distance between the new point and each existing point is considered in large datasets, which lowers the algorithm's speed.
- Because of the multiple levels, the KNN method does not function well with high-dimensional data, making it difficult for the algorithm to calculate the distance in each dimension.
- KNN is sensitive to noise in the dataset. It requires the input of missing values and removal of outliers manually.

8.7.1.6 Random Forest

Random forest or random decision forest is a kind of ML technique for classification, regression, and other tasks. It functions by constructing a number of decision trees at the training time. The output generated by random forest is the class which most trees select for classification task where the mean prediction of the individual trees is considered in case of the regression tasks. One problem of the decision tree is overfitting issue to their training set. Random decision forests correct this. Thus, random forests outperform decision trees. But, their accuracy is not as high as that of gradient boosted trees.

Although random forest is a type of ensemble learning method in principle, its ease of application and frequency of use in different domains have been considered among the basic ML algorithms here. It implements the bagging ensemble learning method, which is described in detail later in this chapter.

Random forests include a particular type of bagging scheme: they employ a modified tree-learning algorithm that selects a subset randomly of the characteristic for each candidate split in the learning process. This process is called "feature bagging." The cause why this is done is the correlation of the ordinary bootstrap sample trees: if one or more functionalities are robust predictors for the response variable or target output, these characteristics will be selected in many of the B trees. These also cause them to become correlated.

Here, about 140 estimators, i.e., decision trees, estimate the final result, resulting in an accuracy of 94.79%. There are several pros and cons of using random forest, which are as follows:

Advantages:

- The base of random forest is bagging algorithm. It also uses the ensemble learning techniques. Many trees are created based on the data subsets, and later, all the outputs of the trees are combined. This leads to reduce the overfitting problem and increase accuracy.
- Random forest requires no feature scaling (standardization and normalization). It uses a rule-based approach in place of distance calculation.
- The performance of random forest is not affected by nonlinear parameters unlike curve-based algorithms. Hence, in case of high nonlinearity between the independent variables, random forest may outperform other curve-based algorithms.
- Random forest can automanage missing values.

Disadvantages:

- Random forest requires creating many trees unlike decision tree. Hence, this algorithm is very demanding in terms of computational power and resources compared to decision tree.
- Random forest is time-consuming; also, it requires a lot of time to train when compared to decision trees as it generates a lot of trees (instead of one tree in case of the decision tree).

8.7.2 Ensemble Learning

The main objective of ensemble models is to improve overall performance of the ML techniques. It works by combining the decisions from multiple models. This chapter focuses on ensemble methods: Max voting, averaging, stacking, bagging, AdaBoost, and gradient boosting. These are discussed hereafter.

8.7.2.1 Max Voting

Max voting is an ensemble learning method. It is used for classification problems as usual. In this method, predictions are made based on multiple models for each data point. When the predictions are made by each model, it is considered as a "vote." The predictions getting the major votes of the models are used as the final prediction. It is a hard voting ensemble where hard voting means predicting the class with the most significant votes from models.

With the chosen dataset, max voting was done using the sklearn VotingClassifier with SVM, regression, Naive Bayes, random forest, decision tree, and KNN as the base prediction models, i.e., the estimators. The accuracy of the max voting ensemble is 96.14%. A comparative study of max voting with the other basic ML methods and the improvements made are given in Table 8.3. Further, the graphical representation of the obtained results and comparative analysis is provided in Figure 8.3.

8.7.2.2 Averaging

Averaging also makes use of the multiple predictions for each data point which is similar to the max voting. Here, to make a final decision, the average is calculated based on predictions of all models. Averaging is a soft voting ensemble method where soft voting predicts the class with the largest summed probability from models. The two properties from artificial neural networks are the main building blocks of averaging.

1. At the cost of increased variance, the bias can be reduced. It will work on any network.
2. If a group of networks is considered, the variance can be reduced at no extra cost to bias. Mathematically, averaging can be expressed as

TABLE 8.3
Comparative Analysis of Elementary Machine-Learning Techniques and Improvement in Accuracy Using Max Voting Technique

Techniques	Accuracy	Change in Accuracy with Max Voting
Decision tree	92.95%	+3.19%
KNN	87.92%	+8.22%
Logistic regression	93.96%	+2.18%
Naive Bayes	92.79%	+3.35%
SVM	93.12%	+3.02%
Random forest	94.79%	+1.35%

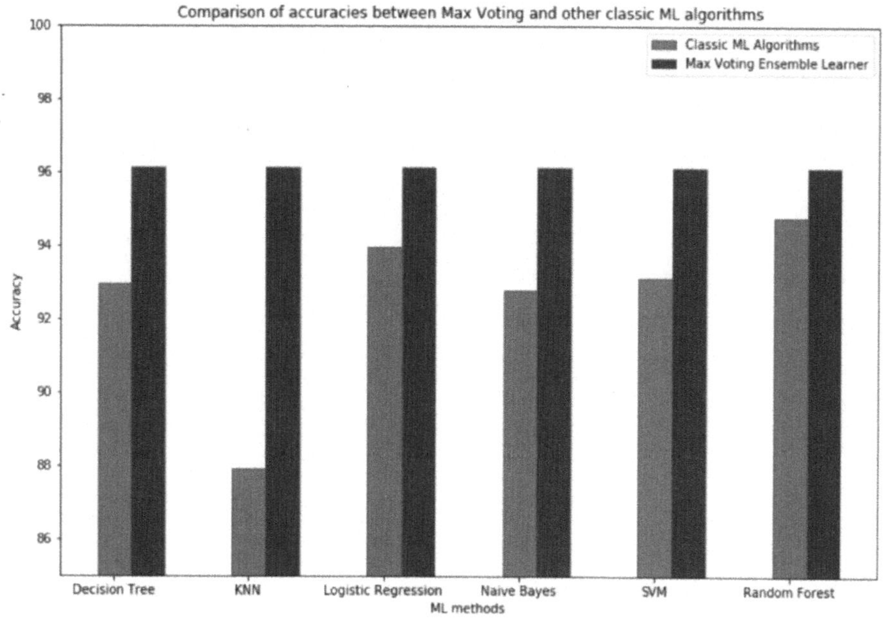

FIGURE 8.3 Comparison of accuracies between max voting and basic machine-learning algorithms.

$$\sum_{i=1}^{N} p_i \, / \, N$$

where p_i is the prediction by the ith model, and N is the total number of models.

Here, the averaging method finds the mean of the predictions done by the six originally used basic machine-learning algorithms—SVM, regression, decision tree, Naive Bayes, KNN, and random forest. The probability of each of these methods is calculated using the predict_proba method. It produces the probability for the target in an array form. The probabilities thus found are then summed and divided by six, the total number of prediction models used here. The averaging method yields an accuracy of 94.97%. A comparative study of averaging with the other basic machine-learning methods and the improvements made are given inTable 8.4. Further, the graphical representation of the obtained results and comparative analysis is provided in Figure 8.4.

8.7.2.3 Stacking

Stacking is an ensemble learning approach where a new model is built using the predictions from multiple models (e.g., decision tree, KNN, or SVM). The model is used to make predictions on the test set. The models (base model) in stacking are typically different and fit the same dataset. A single model (metamodel) is also used

TABLE 8.4
Comparative Analysis of Elementary Machine-Learning Techniques and Improvement in Accuracy Using Averaging Technique

Labels	Accuracy	Change in Accuracy with Averaging
Decision tree	92.95%	+2.02%
KNN	87.92%	+7.05%
Logistic regression	93.96%	+1.01%
Naive Bayes	92.79%	+2.18%
SVM	93.12%	+1.85%
Random forest	94.79%	+0.18%

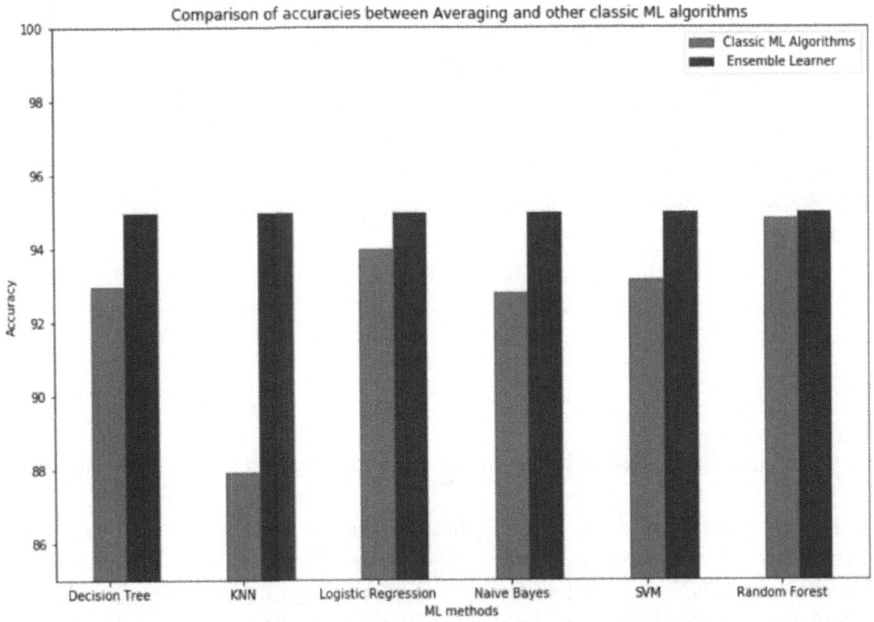

FIGURE 8.4 Comparison of accuracies between averaging and basic machine-learning algorithms.

to study how to best combine the predictions from the contributing models. The steps involved in stacking are explained as follows:

1. The data is first split into two parts, namely, a training set and a test set. Like k-fold cross-validation, the training data is further split into k-folds.
2. A base model (e.g., SVM) is fitted on the $k - 1$ parts. The predictions are made for the kth part.
3. Until every fold is predicted, the process is iterated.
4. The performance of the base model is then calculated. To do so, it is fitted on the entire train dataset.

5. Steps 2to 4 are repeated for other base models (e.g., KNN, decision tree, and logistic regression)
6. To use as the features for the second-level model, predictions from the train set are chosen.
7. The second-level model is used to predict the test set. The outputs generated from the base models are used as the input to the metamodel. They may be actual values if regression is considered, and probability values, probability-like values, or class labels in the case of classification.

Herewith the chosen dataset, the stacking method is used with the base models as—SVM, regression, Naive Bayes, random forest, and KNN. The second-level model used is the decision tree. The meta-model is kept as default, i.e., logistic regression. A comparative study of stacking with the other basic machine-learning methods and the improvements made are given in Table 8.5. Further, the graphical representation of the obtained results and comparative analysis is provided in Figure 8.5.

8.7.2.4 Bagging

The idea lying in bagging is to combine the results of multiple models (like all decision trees) and to get a generalized result. If the predictions of all the models are made on the same set of data, then there is a high chance that all the models will give the same result. So, one solution to this problem is to use bootstrapping. It is a kind of sampling technique. In bootstrapping, from the original dataset, the subset is created based on observations, with replacement. The subset size and original set size are exactly the same. The bagging (or bootstrap aggregating) technique uses these subsets (bags) so that it can get a good idea of the distribution (complete set). The subsets size created for bagging may be less than that of the original set. The steps involved in bagging are as follows:

1. From the original set, multiple subsets are created by selecting the observations with replacement.
2. A base model or weak model is built based on each of the subsets.
3. The models are independent of each other. They also run in parallel.
4. The predictions from all models are combined, and the final predictions are determined.

TABLE 8.5
Comparative Analysis of Elementary Machine-Learning Techniques and Improvement in Accuracy Using the Stacking Technique

Labels	Accuracy	Change in Accuracy with Stacking
Decision tree	92.95%	+2.18%
KNN	87.92%	+7.21%
Logistic regression	93.96%	+1.17%
Naive Bayes	92.79%	+2.34%
SVM	93.12%	+2.01%
Random forest	94.79%	+0.34%

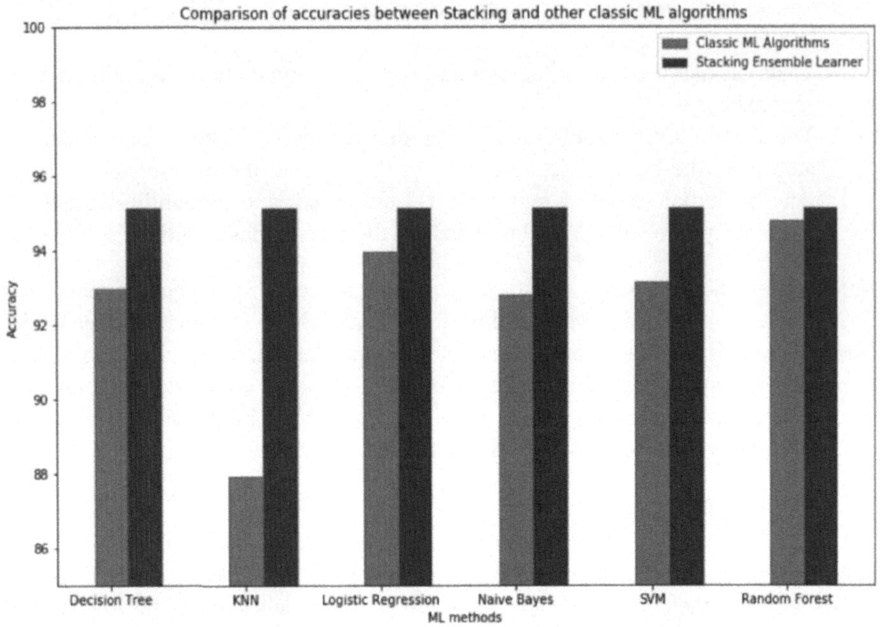

FIGURE 8.5 Comparison of accuracies between stacking and basic machine-learning algorithms.

The method can be implemented in Python using the sklearn bagging classifier which has the following parameters:

- *Base_estimators*: It is used to specify the base estimator to fit on random subsets of the dataset. The default value is a decision tree classifier.
- *N_estimators*: It gives the value of how many estimators are in the ensemble. The value has to be an integer.
- *Max_samples*: It specifies the number of samples to draw from X to train each base estimator.

In the chosen dataset, the base estimator for bagging is kept as a decision tree, giving an accuracy of 96.14%. A comparative study of bagging with the other basic machine-learning methods and the improvements made are given in Table 8.6. Further, the graphical representation of the obtained results and comparative analysis is provided in Figure 8.6.

8.7.2.5 AdaBoost

Boosting is a sequential process. Here, the errors of the previous model are corrected by the attempt of each subsequent model. The succeeding models solely depend on the previous model. Based on a subset of date, a base model is created. Predictions on the whole dataset are made by this model. Actual and predicted values are used

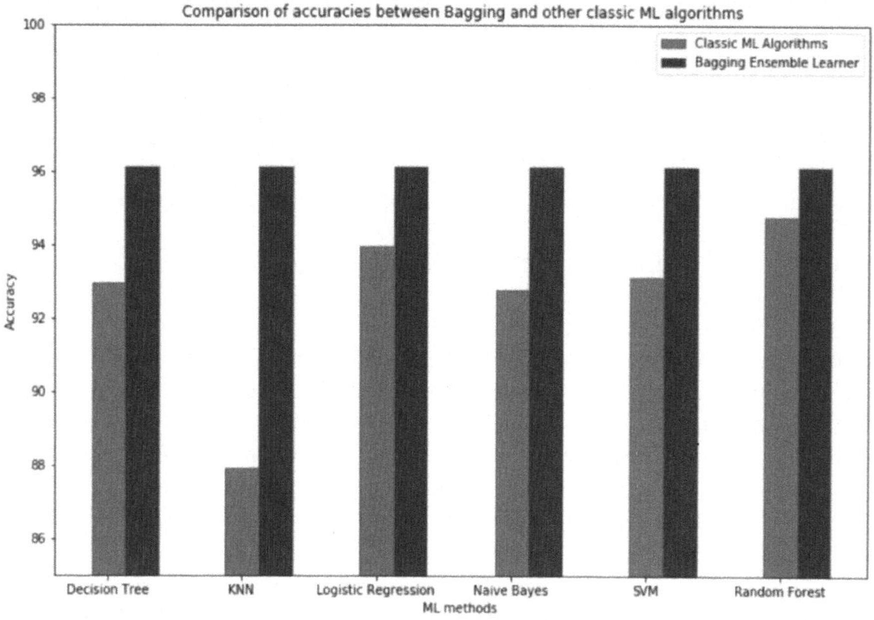

FIGURE 8.6 Comparison of accuracies between bagging and basic machine-learning algorithms.

TABLE 8.6
Comparative Analysis of Elementary Machine-Learning Techniques and Improvement in Accuracy Using Bagging Technique

Labels	Accuracy	Change in Accuracy with Bagging
Decision tree	92.95%	+2.18%
KNN	87.92%	+7.21%
Logistic regression	93.96%	+1.17%
Naive Bayes	92.79%	+2.34%
SVM	93.12%	+2.01%
Random forest	94.79%	+0.34%

to calculate the errors. Incorrectly predicted observations are given higher weights. Another model is created, and predictions are made on the dataset. The new model makes an attempt to correct the errors from the previous model.

In a similar way, multiple models are created, each of which corrects the errors of the previous model. The final model, which is called a strong learner, is the weighted mean of all the weak learner models. In this way, strong learners are formed using weak learners. Though the individual models work well for some parts, they would not work well on the entire dataset. Thus, each model takes part to boost the performance of the ensemble. AdaBoost or adaptive boosting is one of the simplest boosting algorithms. Decision trees are used for modeling usually. Multiple sequential models, which correct the errors of the last model, are created. AdaBoost assigns weights to the incorrectly predicted observations. The subsequent model predicts these values correctly. AdaBoost algorithm steps are given in the following:

1. All observations are given equal weight initially.
2. Based on the subset of data, a model is prepared.
3. Predictions are made on the whole dataset.
4. The predictions and actual values are compared, and the errors are calculated.
5. While creating the next model, the mispredicted data points are given higher weights.
6. The error value is a determining factor of the weights. For instance, the higher the error, the more is the weight assigned to the observation.
7. The process is repeated till the error function changes or the maximum limit of the estimator's number is reached.

For the dataset chosen, the AdaBoost algorithm yields an accuracy of 96.31%. A comparative study of AdaBoost with the other basic machine-learning methods and the improvements made are given in Table 8.7. Further, the graphical representation of the obtained results and comparative analysis is provided in Figure 8.7.

8.7.2.6 Gradient Boosting

Another ensemble learning algorithm is gradient boosting or GBM. It works for both classification and regression problems. Here, boosting technique is used which combines several weak learners to form a strong learner. As base learners, regression trees

TABLE 8.7
Comparative Analysis of Elementary Machine-Learning Techniques and Improvement in Accuracy Using AdaBoost Technique

Labels	Accuracy	Change in Accuracy with AdaBoost
Decision tree	92.95%	+3.36%
KNN	87.92%	+8.39%
Logistic regression	93.96%	+2.35%
Naive Bayes	92.79%	+3.52%
SVM	93.12%	+3.19%
Random forest	94.79%	+1.52%

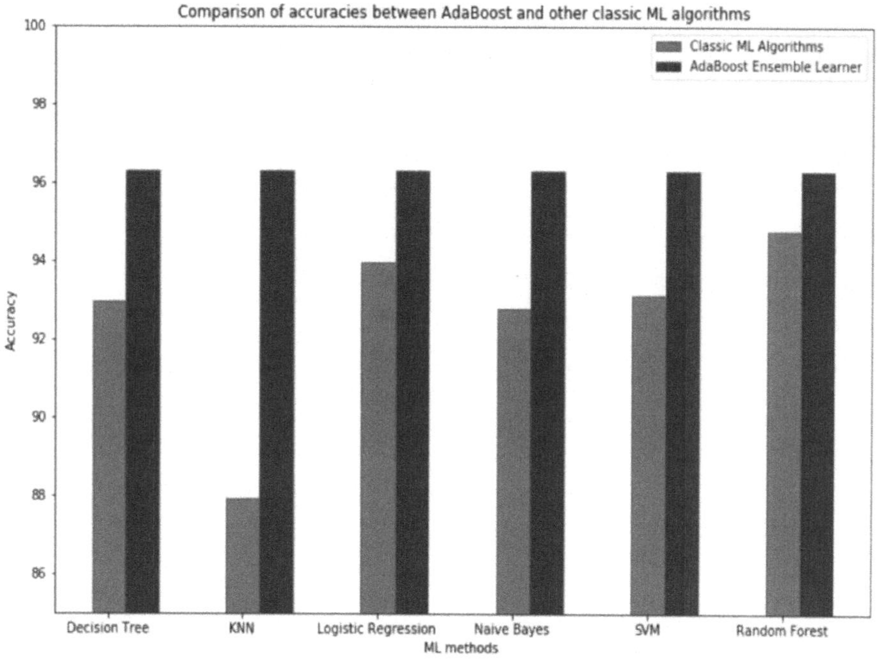

FIGURE 8.7 Comparison of accuracies between AdaBoost and basic machine-learning algorithms.

are used. Based on the errors calculated by the previous tree, each subsequent tree in the series is built. The steps involved in gradient boosting are as follows:

1. The mean value of the target variable is assumed to be the predicted value for all observations in the dataset.
2. This mean prediction and actual values of the target variable are used to calculate the errors.
3. Based on the errors calculated above as the target variable, a tree model is created to find the best split to minimize the errors.
4. The predictions generated by this model are then combined with predictions 1.
5. The calculated value is the new prediction.
6. This predicted value and actual value are now used to calculate errors.
7. Till the maximum number of iterations is reached or convergence is met, steps 2–6 are repeated, *i.e.*, the error function does not change.

The accuracy achieved with gradient boosting is 96.81%. A comparative study of gradient boosting with the other basic machine-learning methods and the improvements made are given in Table 8.8. Further, the graphical representation of the obtained results and comparative analysis is provided in Figure 8.8.

TABLE 8.8
Comparative Analysis of Elementary Machine-Learning Techniques and Improvement in Accuracy Using Gradient Boosting Technique

Labels	Accuracy	Change in Accuracy with Gradient Boosting
Decision tree	92.95%	+3.86%
KNN	87.92%	+8.89%
Logistic regression	93.96%	+2.85%
Naive Bayes	92.79%	+4.02%
SVM	93.12%	+3.69%
Random forest	94.79%	+2.02%

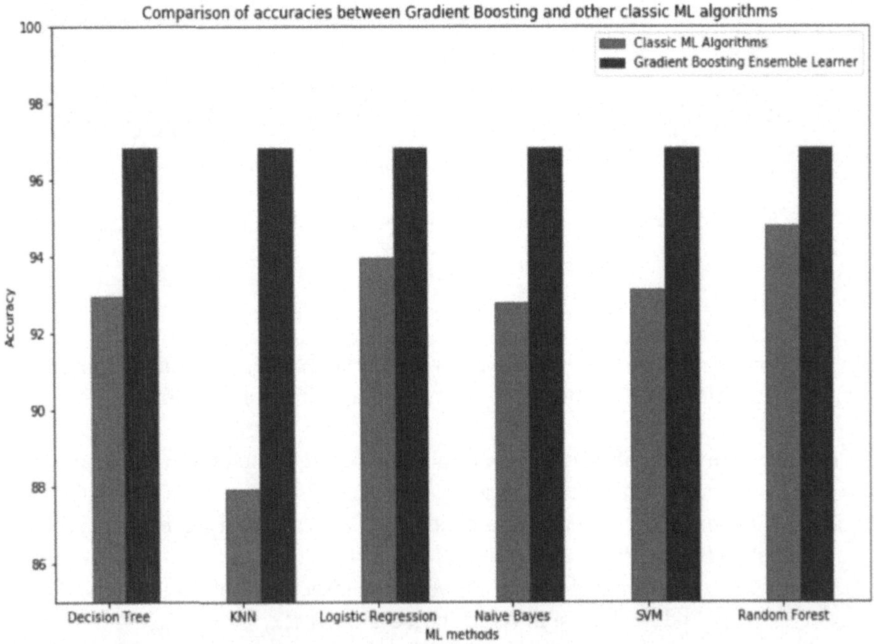

FIGURE 8.8 Comparison of accuracies between gradient boosting and basic machine-learning algorithms.

The implementation of the project has been uploaded to GitHub. GitHub repository link to access and contribute to the project: https://github.com/sasprjiiit/SpearPhishing_Ensemble

8.8 CONCLUSION

From the aforementioned discussion, it is clear that ML is well suited for detecting the patterns found in the tweets. Most of such tweets are used for spreading wrong information or providing baits like offers to the users. Such tweets used an unusual array of words which after training could be adequately identified by learners. SVM and logistic regression performed very well in the ML techniques, providing a brilliant accuracy of about 93%. The ensemble learning techniques, which combined layers of base models of the classical ML algorithms, can enhance or boost the results of the standard classifiers. Although the ensemble learning methods used basic algorithms like SVM, regression, decision tree, KNN, Naïve Bayes for their fundamental estimation, the combination of these methods in different ways, like applying different methods to different sections of the dataset or combining the predictions, introduced diversity in the decision boundary; thus, it improves prediction accuracy. Of the ensemble learning techniques, gradient boosting, in which several weak learners are combined and a strong learner is formed by recalculating and minimizing the error function, provided the best accuracy of about 97%. It is followed by AdaBoost, which reevaluates the mispredicted values and improves upon itself; and the next is max voting, which makes the final prediction based on the decision supported by a majority of the learners, and Bagging, which uses different base estimators independently on a different subset of the dataset for prediction and builds a final prediction by combining these results, respectively. At the end, it can be concluded that the introduction of ensemble methods could provide better results in classifying spear-phishing spam on Twitter than the existing traditional ML approaches.

REFERENCES

Abu-Nimeh, S., D. Nappa, X. Wang, and S. Nair, "A comparison of machine learning techniques for phishing detection." In *eCrime '07: Proceedings of the Anti-phishing Working Groups 2nd Annual eCrime Researchers Summit*. New York, NY: ACM, pp. 60–69, 2007.

Bonaccorso, G., *Machine Learning Algorithms: A Referenceguide to Popular Algorithms for Data Science and Machine Learning*. Packt Publishing, USA, 2017.

Cunningham, P., and S. J. Delany, "K-nearest neighbour classifiers—a tutorial." *ACM Computing Surveys*, vol. 54, no. 6, pp. 1–25, 2021, July.

"Dataset—UtkMl's Twitter Spam Detection Competition." *Kaggle*, 2019 [Online]. Available: www.kaggle.com/c/utkmls-twitter-spam-detection-competition/data

Ferreira, A., L. Coventry, and G. Lenzini, "Principles of Persuasion in social engineering and their use in phishing." In Tryfonas, T., and Askoxylakis, I. (Eds.), *Human Aspects of Information Security, Privacy, and Trust*. Springer International Publishing, USA, pp. 36–47, 2015.

Hosmer, D. W., and S. Lemeshow, *Applied Logistic Regression*. New York: Wiley, 2000.

Mitchell, T. M., *Machine Learning*, 1st ed. New York, NY: McGraw-Hill, Inc., 1997.

Natekin, A., and A. Knoll, "Gradient boosting machines, a tutorial." *Frontiers in Neurorobotics*, vol. 7, p. 21, 2013.

Safavian, S. R., and D. Landgrebe, "A survey of decision tree classifier methodology." *IEEE Transactions on Systems, Man, and Cybernetics*, vol. 21, no. 3, pp. 660–674, 1991, May–June.

Sarkar, D., and V. Natarajan, *Ensemble Machine Learning Cookbook: Over 35 Practical Recipes to Explore Ensemble Machine Learning Techniques Using Python*. Packt Publishing, 2019. [Online]. Available: https://books.google.co.in/books?id=dCWGDwAAQBAJ

Seymour, J., and P. Tully, "Weaponizing data science for social engineering: Automated E2E spear phishing on Twitter." 2016. https://av.tib.eu/media/36231 (Accessed on 15 September 2020).

Vergelis, M., T. Shcherbakova, and T. Sidorina, “Spam and phishing in Q1 2019” [Online], 2019. Available: https://securelist.com/spam-and-phishing-in-q1-2019/90795. (Accessed on 17 October 2020).

Vishwanath, A., T. Herath, R. Chen, J. Wang, and H. R. Rao, “Why do people get phished? Testing individual differences in phishing vulnerability within an integrated information processing model.” *Decision Support Systems*, vol. 51, pp. 576–586, 2011.

Yeboah-Boateng, Ezer Osei, and Priscilla Mateko Amanor, “Phishing SMiShing & Vishing: An assessment of threats against mobile devices.” *Journal of Emerging Trends in Computing and Information Sciences*, vol. 5, no. 4, pp. 297–307, 2014, April.

9 RFID and Operational Performances

A Prospect of Inventory Management Practice in Medical Store

Duryodhan Jena, Rashmi Ranjan Panigrahi, Meenakshee Dash, and Anil Kumar Bhuyan

CONTENTS

9.1 INTRODUCTION

In spite of the fact that there is well-documented evidence of improvements for significant competitiveness and reduced costs by incorporating inventory management (SCM), the healthcare industry has been exceedingly hesitant to adopt these practices (Regattieri et al. 2018; Panigrahi et al. 2021c).

DOI: 10.1201/9781003152392-9

The aim of healthcare systems is to achieve maximum patient care; even so, the ineffectiveness of productive areas can have substantial effects on the quality of services. The hospital is more than simply a connected client, and, as a result of its numerous and changeable material characteristics, its institutional supply chain is massively complicated. Avoiding waste and enhancing efficiency are, therefore, a global problem in this complex system, emphasizing that any priority action source needs to be identified and any methodology and technology used to increase healthcare provision and services. Drug logistics plays a key role in this scenario because the related acquisitions and management expenses have a major effect on cost (Panigrahi et al. 2020).

On the basis of literature, the methods used most often to distribute supplies to the hospital facility over the decades at least changed from clinical practice demand-based processes, transaction carts, and recurring automatic refurbish or par level systems to the unique two-bin system, radio frequency identification two-bin system, weight reduction garbage cans, and user-driven unit systems (Almutairi et al. 2019). RFID is recognized as an energy-efficient practice of inventory management. It consumes less energy as compared to other practices. It is frequently used as an energy-efficient technique because it easily records the data as long as it is in its periphery or reader vicinity. This technique is widely used in the supply chain management area. Radio frequency identification practices in inventory were considered to be the most effective mechanism for every commercial business (Velasco et al. 2018). To cope with the rapid change of technology, we should have control over assets tracking and warehouses management systems. By looking at the two control systems mentioned earlier, RFID is being adopted by retail outlets (Prachař et al. 2014). From the point of view of SML, RFID (global leader for providing high-performance RFID tech. solution for retail) explains that RFID market continues to grow by 30% year after year. By the use of RFID practices, inventory accuracy improved by 98% and more as compared to other stock management techniques. This chapter also gives importance to the popularity and wide acceptance of RFID practices in SCM. Subsequently, it reduces the time element for repetitious work. Report of IDTechEx. reveals that the value of RFID has been increased from 11.6 billion to 13 billion by the end of 2022. RFID includes tags, readers, labels, software cum service, cards, and other items of radio frequencies along with passive and active RFID (Surana et al. 2005). This technique will be less costly as compared to printed codes. RFID and barcode are two accepted technologies for auto identification products in stores. In recent times, RFID practices are applied to increase the efficiency of operation (Guarte et al. 2006). In warehouse management, the most used technology is RFID. Indian retails firms widely used RFID practices, and it helps compete with global market. Still, there is gap in understanding RFID practices in the medium and large retail in healthcare sector in India (Camdereli et al. 2010).

There are numerous different drugs management systems in the real world, mainly because of the existence of certain conditions or specific managers with management approaches. However, a systematic approach is missing which can help the process of improvement. In this context, the authors suggest a unique IMP approach for introducing RFIPIMP (radio frequencies identification practices inventory management practices) concepts into a drug supply chain. It was tested in an important hospital in Odisha. To encourage future course of action that can help Odisha medical organizations achieve their mission, an assessment of current practices is required and to guide future research in the sector (Qiao et al. 2013). The study tries to give importance on the present RFID (automated inventory practices) adopted in the medium and large retail pharmaceutical stores and looks at how it contributes to store performances. Section 9.2 gives a brief literature of review, relevant theory building related to healthcare industry, together with essential topics in the field of healthcare logistics. Section 9.3 explains about **energy-efficient RFID practices in medical store inside hospitals**, Section 9.4 describes research gap, objectives, and proposed hypothesis of the study. Section 9.5 describes the study technique and methodology employed in the collection and analysis of the data. Section 9.6 explains the result and discusses the effectiveness of RFID as IM practice for smooth store operation, followed by Section 9.7 provides concluding remark and area of further research.

9.2 REVIEW OF LITERATURE

The ongoing drive to reduce cost and better manage their activities continues to confront hospitals throughout the world while satisfying the requirements of a "changing demands" society. Supply chains account for up to 46% of healthcare costs as just a significant component of healthcare costs (Nayak et al. 2015). Budget for healthcare potential savings and effective means to improve hospital services can be a key contributor to the management of inventory facilities (Jack et al. 2009).

RFID is one of the most vital and prevalent systems, as per previous research, with connections between both industry and potential health. Techniques or processes for energy savings are regularly attempted by the business or medical stores inside hospitals. In a business that minimizes the operating costs of various systems, RFID is an energy-saving strategy. When energy efficiency decreases operational costs, the rate of return automatically increases. RFID strategies minimize retail energy consumption but instead contribute to smooth functioning. Also, they help avoid massive costs (Böhme et al. 2013) (Krejcie et al. 1970).

RFID technique is used in each business to constantly capture material things or equipment through radio signals. RFID techniques are used to enhance transactional movement and identification in the retail units of healthcare industry. This technique, identified as the most rapidly increasing wireless technology, has a significant impact on operational cost retail store. By using the communication tags, RFID offers retail sector more millage to monitor business with maximum precision (Li et al. 2017).

9.2.1 Theoretical Foundation

We used two theories in this essay to create the foundation. It was noted from the earlier studies that the utilization of technological progress in any industry most of the time suggests how people accept the notion. Rogers' theories and business information technology are used with two theories. Energy model explains RFID techniques to produce energy consumption (Patil et al. 2007).

9.2.2 First Energy Model

To check the energy efficiency, we have to adopt the energy model, which is based on the size of data used (recorded in a bit), rate of recording (recorded in kbps), time consumed to send the data, and lastly power supply (Jena et al. 2020).

Model can be explained as

E = Energy, V = Volt used in power supply, I = amperes for consumption of current at the time of scanning. In the prospects of energy consumption, RFID has three modes of operation: 1. scan, 2. idle, and 3. sleep, Out of two modes, first mode consumes the highest energy, and the last mode consumes the least energy. Due to repetitions of work, sleep mode is frequently used in a firm or industry.

9.2.3 Second Rogers' Theory of Diffusion of Innovation

Rogers' decision to adopt and use basically took place in between five interlinked stages. The five stages of decision-making of knowledge, persuasion, decision, implementation, and confirmation are interlinked, and at the time of taking any kind of decision, they play a significant role in an organization (Anjum et al. 2019). Though the theory has been introduced by Rogers in 1995, still today, it creates an impact on industries. Along with the five stages, it has been supported by five technological attributes, which help an organization's smooth adoption of technological innovation. The relative advantages are compatibility, complexity, observability, and trialability. Among these five attributed most accepted attributes used in technology adoption as per researcher point of view are relative advantages, compatibility, and complexity (Bhiradi et al. 2014).

9.3 ENERGY-EFFICIENT RFID PRACTICES IN MEDICAL STORE INSIDE HOSPITALS

Radiofrequency practices provide an extraordinary opportunity to enhance shopping experiences to the customer as well as provide new ways to adopt different offerings in a single platform. This facility with minimum effort of energy consumption may give extra millage to retailers (Chanchaichujit et al. 2020) (Muhammad-Masum et al. 2013).

9.3.1 Operational performance of stores improved through RFID (as an energy-efficient technique)

Performances of the store improve if the following points are considered and adopted prior to the introduction of RFID practices. The systems mentioned in the following list need to be installed in advance to RFID installed.

- RFID labeling station
- RFID reader for inventorying the store
- Overhead antennas for real-time inventory of the warehouse.
- RFID point of sale
- RFID-enabled floor mat EAS

As per Gary Lynch FCILT (CEO GS1 UK), RFID was evolved as multisolution provider in the area of inventory management in retail sector. These areas are improved IM, improving performance of sales and delivery to omnichannel processes.

9.4 RESEARCH GAP, OBJECTIVES, AND HYPOTHESIS

9.4.1 Research Gap

Existing literature shows that in medium-to-large retail companies, RFID plays a major role. Nowadays, although, the large retail sector relies heavily on RFID practices. Many articles explain the energy consumption aspect of RFID and its relation to management strategies of inventory (Thanapal et al. 2017). But in Indian contexts, the link between energy-efficient RFID practice and business performance in retail shops was not investigated. We have, therefore, carried out the study to discover how energy-efficient RFID can affect medical store performance in the Indian retail industry as an inventory practice (Guide et al. 2020).

9.4.2 Research Objective and Hypothesis

RFID practices are considered to be in their infancy stage, but they have a positive impact on automated inventory management practices (Victoire et al. 2015). So, in this chapter, an attempt has been made to check the impact of RFID practices toward increasing the operational performance of retail stores. In accordance with above literature and our study objectives, this chapter proposed the hypothesis to being tested in an Indian scenario with respect to the state of Odisha. The proposed hypothesis will be: H1 = RFID; it will have a significant impact on the performance of medical stores inside hospitals.

9.5 METHODOLOGICAL FOUNDATION

The researcher applies PCA to confirm critical components which contribute heavily to the operating performance of the store and to the latent factor RFIDs. After this, the connection between RFID and store performance is been examined using correlation. Data were collected via Google form and online mode from top 15 government of Odisha registered private hospital's medical stores. Fifteen respondents have been considered under each medical store as a sample unit by utilizing convenience sampling technique. This current study comprises a total population of 1,500 employees. Participants were selected from 15private medical stores of hospitals registered under government of Odisha, i.e., Kalinga Hospitals, Aditya Care Hospitals, L.V. Prasad Eye Institute, Ashwini Hospital, Ayush Hospital, Apollo Hospital, Kanungo Institute of Diabetic Specialities, Kalinga Institute of Medical Sciences, Hi-Tech Medical College & Hospital, IMS & SUM Hospital, M/s Shanti Memorial Hospital, Chitta Ranjan Seva Sadan, Christian Hospital, Christian Hospital, and CARE Hospital by using convenience sampling (Boyinbode et al. 2015). Exactly 194 relevant questionnaires (86%) have been considered for analysis out of 225 distributed sample questionnaires. The method was established by Krejcie and Morgan to determine the required sample size (38) for a finite population. The minimum sample size was calculated via the estimation process 172. The validity and dependability of data were measured by the pilot test. Cronbach's alpha values were greater than 0.81, indicating that the study's variables were considered reliable. The respondents were key officials of the medical store inside hospitals, i.e., store manager, operations manager, purchase manager, and warehouse manager. A five-point Likert scale (where 1—not at all effective, 2—not effective, 3—somewhat at, 4—effective, and 5—very effective) used as valid scale of measuring instrument and data collection. Both RFIP (Radio Frequency inventory Practices) and OP (Operational performance) was measured through five-point Likert scale.

9.6 RESULTS AND ANALYSIS

9.6.1 Principal Component Analysis

In this research, we have employed PCA to check key items (questions) and decrease the items of less than 1 own value and value based on RFID practices. After PCA with rotation of varimax of 30 items, 17 items have been selected based on criteria of minimum threshold limit of KMO value 0.60 and more, communalities value 0.50 value, and lastly also checked minimum fulfilling criteria measures of sampling adequacy that is more than the KMO value of anti-image matrices of 0.60. By applying the aforementioned three threshold limits, after checking with above mentioned parameter's thresholds limit seven items of RFID is selected out of thirteen items, and ten items of improved medical store performance is selected out of seventeen items (refer Table 9.1).

9.6.1.1 Correlation Analysis

H1 = RFID has a significant impact on the performance of medical stores inside hospitals—0.901

This section states the correlation between the latent variables, that is, RFID with store operational performance. This analysis reveals that the subcomponent of RFID has a significant and positive relationship with store operational performance. This study justifies the statement with a correlation value of 0.911 (RFID) with store operational performance. The output does not deviate from previous studies (5,34,39–41) (Table 9.2).

TABLE 9.1
Factor Analysis by Using (Principal Component Analysis) of RFID and Store Performances

Rotated Component Matrix

Items under Store operating performance and energy economy RFID	Component	
	1	2
RFIP 6—Enhanced distribution systems for warehouses	.944	
RFIP 4—-Encourages partners to meet intended audience demand through distribution networks	.941	
RFIP 8—Easy replace/return items tracked in store	.911	
RFIP 1—Revenue generation through rapid transformation of the firm with less power consumption	.766	
RFIP 5—Inventory practices help minimize physical store overstock issues	.759	
RFIP 9—Inventory practices help minimize physical store overstock issues	.731	
RFIP 3—Provides competitive benefits	.722	

SOP 9—Brings reliability to storing, in an energy-efficient way	.933
SOP 7—Theft/fraud security control as well as other illicit operations	.914
SOP 1—-Fehler reduction in stock management records at retail establishments	.901
SOP 10—Enhancing performance through IT	.872
SOP 6—Lack of sales-driven automated inventory processes	.854
SOP 8—Reduced overall energy-efficient storage cost	.834
SOP 2—Improved handling of inventories	.824
SOP 4—Improved handling of inventories	.811
SOP 5—Fewerdamagesrecordedingodown	.802
SOP 3—Due to accuracy in information decision-making easier, it is given effective result.	.712

Source: Author's computation (2021)

TABLE 9.2
Pearson's Correlations in between RFID with Store Performance

Correlations

		RF	STP	RFIP 1	RFIP 3	RFIP 4	RFIP 5	RFIP 6	RFIP 8	RFIP 9
RFID practices	Pearson correlation	1	.911**	.855**	.527**	.812**	.920**	.841**	.866**	.749**
Store performance	Pearson correlation		1	.857**	.405**	.681**	.854**	.762**	.768**	.752**
RFIP 1	Pearson correlation			1	.327**	.669**	.843**	.683**	.693**	.574**
RFIP 3	Pearson correlation				1	.075	.466**	.162	.162	.717**
RFIP 4	Pearson correlation					1	.722**	.816**	.906**	.338**
RFIP 5	Pearson correlation						1	.713**	.748**	.680**
RFIP 6	Pearson correlation							1	.895**	.412**
RFIP 8	Pearson correlation								1	.420**
RFIP 9	Pearson correlation									1

**. *Correlation is significant at the 0.01 level (2-tailed).*

Source: Author's computation (2021)

TABLE 9.3
Model Summary Regression between Two Automation Inventory Practices (RFID) versus Store Performance

Model Summary

Model	R	R Square	Adjusted R Square	Std. Error of the Estimate	Change Statistics					Durbin–Watson
					R Square Change	F Change	df1	df2	Sig. F Change	
1	.941[a]	.901	.893	.314	.901	137.074	7	102	.000	1.76

a. Predictors: (Constant), barcode practices, radio-frequency practices.

Source: Author's computation (2021)

9.6.1.2 Regression Analysis

In this way, R square and R are interpreted where R square refers to the quantity of DV (dependent variable—storage operating performance) variable in the model series IV (Independent variable—RFID). The R square value is 0.901, which shows that the performance of the storage area is 90% dependent on RFID technology. R can explain the correspondence coefficient of 94.1% between DV and IV. According to the D-W values, close to 2 indicates regarding non-autocorrelation in the regression model. So, as per calculation, the study depicts it is free from (D-W - 1.76) autocorrelation (Table 9.3).

9.7 CONCLUSION AND SCOPE FOR FURTHER RESEARCH

This chapter focuses on the impact on hospital operational performance of efficient RFID procedures in the medical stores. The RFID practice is mainly concerned with the operating performance of storage in the area. Inventory practices help avoid stocks in retail outlets, i.e., facilitate the fulfillment of customer demand in distribution canals, improved warehouse distribution systems, and ease of replacement/return storage tracking. The study not only concludes on the basis of relationships but also offers retail officers' significant feedback on automated inventory management system adoption. It tracks the lost things simply and restricts the internal theft and misuse/maltreatment through the appropriate tracking mechanism. Bypassing the time store, the upgrade must be brought into operation, which, together with investment made in the inventory, can control a higher number of satisfied customers. Great amount of money is not blocked by stocks in godowns. The study's contribution to society and the retail industry is to save time, time, and effort by using these techniques and to optimize resources, and finally to provide a services industry that needs to know whether retailers are adopting centralized systems or a decentralized system in store operations, according to the organization's size. RFID practice is operating at the medical stores within hospitals thanks to the use of the energy-efficient technology.

Furthermore, with larger samples and broader coverage and sample medical store, more than 15 hospital units can be considered (including government medical units) and applied, providing a convincing result. Subjective assessment might lead to a relatively modest reliability increase in measurement error. Objective measurement will provide a better analysis with more accurate information. Knowledge of stock automation procedures has been looked at to close the gap between concept and practice. This area can be emphasized and broadened in future studies.

Ethical Approval

Authors' Contribution: Duryodhan Jena, Rashmi Ranjan Panigrahi, Meenakshee Dash, and Anil Kumar Bhuyan contributed to the design and implementation of the research to the analysis of results. All the authors read and finally approved the paper.

Conflict of Interest: The authors certify that they have no affiliations with or involvement in any organization or entity with any financial interest or nonfinancial interest in the subject matter or materials discussed in this manuscript.

Funding Acknowledgment: The authors declare that they have not received any financial support for the research, authorship, and/or the publication of this book.

REFERENCES

Almutairi AM, Salonitis K, Al-Ashaab A. Assessing the leanness of a supply chain using multi-grade fuzzy logic: a health-care case study. *Int J Lean Six Sigma.* 2019;10(1):81–105.

Anjum SS, Noor RM, Anisi MH, Ahmedy IB, Othman F, Alam M, et al. Energy management in RFID-Sensor Networks: Taxonomy and challenges. *IEEE Internet Things J.* 2019;6(1):250–266.

Bhiradi I, Pillai JA. Tool Inventory Management Using RFID Technology. *5th Int 26th All India ManufTechnolDesResConf*[Internet].2014;12:14–17.Available from: www.researchgate.net/publication/329442493_tool_inventory_management_using_rfid_technology

Böhme T, Williams SJ, Childerhouse P, Deakins E, Towill D. Methodology challenges associated with benchmarking healthcare supply chains. *Prod Plan Control.* 2013;24(10–11):1002–1014.

Boyinbode O, Akinyede O. A RFID based inventory control system for Nigerian supermarkets. *Int J Comput Appl.* 2015;116(7):7–12.

Camdereli AZ, Swaminathan JM. Misplaced inventory and radio-frequency identification (RFID) technology: information and coordination. *Prod Oper Manag.* 2010;19(1):1–18.

Chanchaichujit J, Balasubramanian S, Charmaine NSM. A systematic literature review on the benefit-drivers of RFID implementation in supply chains and its impact on organizational competitive advantage. *Cogent Bus Manag* [Internet]. 2020;7(1). https://doi.org/10.1080/23311975.2020.1818408

Guarte JM, Barrios EB. Estimation under purposive sampling. *Commun Stat Simul Comput.* 2006;35(2):277–284.

Guide SL. Effectively managing inventory & operations through automation [Internet]. *Stitch Labs 2019.* 2019 [cited 2020 Jul 2]. Available from: www.stitchlabs.com/resource/guides/effectively-managing-inventory-and-operations-through-automation/

Jack EP, Powers TL. A review and synthesis of demand management, capacity management and performance in health-care services. *Int J Manag Rev.* 2009;11(2):149–174.

Jena D, Panigrahi RR, Gangwar VP, Bhuyan AK. Role of rfid on store operational performances—A prospect of inventory management practice. *Int J Psychosoc Rehabil.* 2020;24(6):10014–10024.

Krejcie RV, Morgan D. Determining sample size for research activities- samlpe techniques. *NEA Res Bull.* 1970;30:607–610.

Li Z, Liu G, Liu L, Lai X, Xu G. IoT-based tracking and tracing platform for prepackaged food supply chain. *Ind Manag Data Syst.* 2017;117(9):1906–1916.

Muhammad-Masum AK, Bhuiyan F, Kalam-Azad A. Impact of radio frequency identification (RFID) technology on supply chain efficiency, an extensive study. *Glob J Res Eng Civ Struct Eng.* 2013;13(4).

Nayak R, Singh A, Padhye R, Wang L. RFID in textile and clothing manufacturing: technology and challenges. *Fash Text.* 2015;2(1).

Panigrahi RR, Jena D. Inventory control for materials management functions—a conceptual study. In: Patnaik S, Ip A, Tavana MJV, editor. *New Paradigm in Decision Science and Management Advances in Intelligent Systems and Computing. Proceeding.* Singapore: Springer Nature Singapore Pte Ltd.; 2020. p. 187–193.

Panigrahi RR, Jena D, Jena A. Deployment of RFID technology in Steel Manufacturing Industry—An inventory management prospective. In: Patnaik, Srikanta X-SYKS, editor. *Advances in Machine Learning and Computational Intelligence Algorithms for Intelligent Systems Springer, Singapore* [Internet]. Singapore: Springer, Singapore; 2021a. p. 705–719. Available from: https://link.springer.com/chapter/10.1007/978-981-15-5243-4_67

Panigrahi RR, Jena D, Mishra PC. Inventory automation practices and productivity: a study on steel manufacturing firms. *Int J Appl Syst Stud.* 2021b;X(xxxx):1–19.

Panigrahi RR, Jena D, Tandon D, Meher JR, Mishra PC, Sahoo A. Inventory management and performance of manufacturing firms. *Int J Value Chain Manag.* 2021c;12(2):149–170.

Patil H, Chaudhari M, Gore A, Kale R, Hingoliwala A, Professor A. A survey on technologies used for billing system in supermarkets. *Int J Innov Res Sci Eng Technol (An ISO Certif Organ).* 2007;3297(10):9913–9918.

Prachař J, Fidlerová H, Sakál P, Zbojová T. Improving the sustainability and effectiveness of the inventory management in manufacturing company. *Appl Mech Mater.* 2014;693(February 2019):141–146.

Qiao Y, Chen S, Li T. Tag-ordering polling protocols in RFID systems. *SpringerBriefs Comput Sci.* 2013;(9781461452294):59–82.

Regattieri A, Bartolini A, Cima M, Fanti MG, Lauritano D. An innovative procedure for introducing the lean concept into the internal drug supply chain of a hospital. *TQM J.* 2018;30(6):717–731.

Surana A, Kumara S, Greaves M, Raghavan UN. Supply-chain networks: a complex adaptive systems perspective. *Int J Prod Res.* 2005;43:4235–4265.

Thanapal P, Prabhu J, Jakhar M. A survey on barcode RFID and NFC. *IOP Conf Ser Mater Sci Eng.* 2017;263(4):0–9.

Velasco N, Moreno JP, Rebolledo C. Logistics practices in healthcare organizations in Bogota.*Acad Rev Latinoam Adm.* 2018;31(3):519–533.

Victoire M. Inventory management techniques and its contribution on better management of manufacturing companies in RWANDA case study: SULFO RWANDA Ltd. *Eur J Acad Essays ISSN* [Internet]. 2015;2(6 Online):49–58. Available from: www.euroessays.org

Index

A

B

C

D

E

F

G

H

I

J

K

L